ELECTRICITY

HOW TO ORDER THIS BOOK

BY PHONE: 800-233-9936 or 717-291-5609, 8AM–5PM Eastern Time

BY FAX: 717-295-4538

BY MAIL: Order Department
Technomic Publishing Company, Inc.
851 New Holland Avenue, Box 3535
Lancaster, PA 17604, U.S.A.

BY CREDIT CARD: American Express, VISA, MasterCard

BY WWW SITE: http://www.techpub.com

FUNDAMENTALS *for the* WATER & WASTEWATER
MAINTENANCE OPERATOR SERIES

ELECTRICITY

FRANK R. SPELLMAN, Ph.D.
JOANNE DRINAN

TECHNOMIC
PUBLISHING CO., INC.
LANCASTER • BASEL

Electricity
a**TECHNOMIC**® publication

Technomic Publishing Company, Inc.
851 New Holland Avenue, Box 3535
Lancaster, Pennsylvania 17604 U.S.A.

Printed in the United States of America
10 9 8 7 6 5 4 3 2 1

Main entry under title:
Fundamentals for the Water & Wastewater Maintenance Operator Series: Electricity

A Technomic Publishing Company book
Bibliography: p.
Includes index p. 251

Library of Congress Catalog Card No. 00-107617
ISBN No. 1-56676-907-8

Contents

3 D-C CIRCUITS

4 CONDUCTORS

8 CAPACITANCE

9 INDUCTIVE AND CAPACITIVE REACTANCE

10 A-C CIRCUIT THEORY

Series Preface

Currently several books address broad areas of wastewater and waterworks operation. Persons seeking information for professional development in water and wastewater can locate study guides and also find materials on technical processes such as activated sludge, screening and coagulation. What have not been available until now are accessible treatments of each of the numerous specialty areas that operators must master to perform plant maintenance activities and at the same time to upgrade their knowledge and skills for higher levels of certification.

The *Fundamentals for the Water and Wastewater Maintenance Operator Series* is designed to meet the needs of operators who require essential background knowledge of subjects often overlooked or covered superficially in other sources. Written specifically for maintenance operators, the series comprises focused books designed to enhance knowledge and understanding.

Fundamentals for the Water and Wastewater Maintenance Operator Series covers over a dozen subjects in volumes that form stand-alone information guides or elements of a library of key topics. Areas to be presented in series volumes include: electricity, electronics, water hydraulics, water pumps, handtools, blueprint reading, piping systems, lubrication, and data collection.

Each volume in the series is written in a straightforward style without jargon or complex calculations. All are heavily illustrated and include extensive, clearly outlined sample problems. Self-check tests are found within every chapter, and a comprehensive examination concludes each book.

The series provides operators with the information required for improved job performance. Equally important, using key points, worked problems, and sample test questions, the series is designed to help operators answer questions and solve problems on certification and licensure examinations.

Preface

Without a doubt, imagining our modern world without electricity would be difficult. More difficult would be for water/wastewater operators to imagine their treatment facility without electricity. Electricity pervades our lives at every turn and its applications are myriad. It has made possible an almost infinite variety of tools, machinery and other equipment, most of which we take for granted—from the simple flashlight to incredibly complex computerized systems that operate many of our plants. We have made ourselves almost completely dependent on it.

This text is not primarily concerned with applications, but with the basic principles of electricity underlying all the applications. It will assist you in gaining an understanding of the laws governing voltage, current, resistance, and power, as well as the somewhat more complex concepts of alternating current—the principles that are the basis for electronics. The goal of this text is not to make electricians of anyone but to provide basic electrical information for maintenance operators in both water and wastewater treatment and thus give them a better understanding of the complicated equipment found in plants.

A knowledge of the simple equations of basic mathematics is necessary for the study of electricity and electronics. However, to make your journey through this text less complicated we present many example problems in simple step-by-step format. Explanations and solutions are sufficiently detailed that the text can stand alone, making it useful for self-study and reference.

To assure correlation to modern practice and design, we present illustrative problems in terms of commonly used voltages and current ratings, and cover circuits and equipment typical of those found in today's electrical systems.

Each chapter ends with a self-test to help you evaluate your mastery of the concepts presented. Before going on to the next chapter, take the self-test, compare your answers to the key, and review the pertinent information for any problems you missed. If you miss many items, review the whole chapter. A comprehensive final exam can be found at the end of this book.

IMPORTANT ***Note:*** *This symbol (to the left) displayed in various locations throughout the manual indicates or emphasizes an important point or points which the operator should read carefully.*

The text may be used by those who have no experience with electricity. If you work through the text systematically, you will be surprised at how easily you acquire an understanding of the fundamentals of electricity.

Acknowledgments

Water/wastewater workers throughout the world have worked with, around, and (in some cases) on electrical machinery of all kinds. Waterworks operators in particular must know basic electrical principles and concepts for their jobs, and many licensure examinations test them on this information. Unfortunately, training in basic electricity has not always been available or afforded to operators who needed it. This is no longer the case, because this manual fulfills the need. In several other volumes on other maintenance topics to follow, we'll address other information operators also need to know.

We have stated the purpose, and now we state the objective. Simply put, this manual's objective is to bring together under one cover the most salient basic theory based on principles, practices, and processes of a sound basic electricity training program. This manual also provides a ready source of practical answers to the most common electrical problems facing those water/wastewater operators who practice, teach, and learn as well as for those who intend to practice, teach, or learn basic electricity.

The water/wastewater professionals with whom we are in constant contact in our careers as authors, lecturers, trainers, seminar presenters, and consultants have been very generous in sharing their expertise and experiences with us. The problems they have encountered and solutions that worked have been a rich source of useful information. To all of them, we continue to owe a large debt of gratitude.

In particular, we acknowledge the extensive training given to us by the U.S. Navy. Both of us had the distinct honor and pleasure of having served extended active duty careers in the Naval Service, where we had the good fortune of being trained at the Navy's Class A (basic), B (advanced), and several C (very advanced) schools. We agree that the classroom training and subsequent on-the-job-training we received while on active duty was far superior to any that we could have received elsewhere. For this, we are eternally grateful.

This manual (and many of the volumes to follow) is a direct result of the extensive training we received while in the Navy. In particular, however, we need to acknowledge the help provided us by the main reference resource we used in compiling this particular manual: The Bureau of Naval Personnel's *Basic Electricity* (1970), available through Dover Publications, Inc., New York, NY.

Electricity: What Is It?

TOPICS

Nature of Electricity
The Structure of Matter
Conductors, Semiconductors, and Insulators
Static Electricity
Magnetism
Difference in Potential
Electric Current
Resistance

1.1 INTRODUCTION

Water/wastewater operators generally have little difficulty in recognizing electrical equipment. Electrical equipment is everywhere and is easy to spot. For example, typical plant sites are outfitted with electrical equipment that

- generates electricity (a generator or emergency generator)
- stores electricity (batteries)
- changes electricity from one form to another (transformers)
- transports or transmits and distributes electricity throughout the plant site (wiring distribution systems)
- measures electricity (meters)
- converts electricity into other forms of energy (rotating shafts—mechanical energy, heat energy, light energy, chemical energy, or radio energy)
- protects other electrical equipment (fuses, circuit breakers, or relays)
- operates and controls other electrical equipment (motor controllers)

- converts some condition or occurrence into an electric signal (sensors)
- converts some measured variable to a representative electrical signal (transducers or transmitters)

Key Terms Used in This Chapter

MOLECULE	Two or more atoms joined together by a chemical bond.
ELECTRON	A tiny, negatively charged particle that orbits around the nucleus of an atom.
PROTON	A positively charged particle located in the nucleus of an atom. The number of protons in the nucleus of an atom determines the atomic number of that element.
NEUTRON	An uncharged elementary particle that has a mass approximately equal to that of the proton.
ION	An atom is electrically unstable because it has more or fewer electrons than protons. A positive ion is called a cation. A negative ion is called an anion.
CONDUCTOR	A substance that permits the flow of electricity, especially one that conducts electricity with ease.
INSULATOR	A substance that offers very great resistance, or hindrance, to the flow of electric current.
RESISTANCE	Opposition to current flow in an electric circuit.
CURRENT	The flow rate of electricity, measured in amperes (amps).
DIFFERENCE IN POTENTIAL	The pressure of electricity, measured in volts.
PERMEABILITY	The relative ease with which a substance conducts magnetic lines of force.
CONDUCTANCE	The ability of a material to pass electrons.

Recognizing electrical equipment is easy because we use so much of it. If we ask typical water/wastewater maintenance operators where such equipment is located in their plant sites, they know, because they probably perform maintenance on the same devices or their ancillaries. If we asked these same maintenance operators what a particular electrical device does, they could probably tell us. If we were to ask if their plant electrical equipment was important to plant operations, the chorus would resound, "absolutely."

Here's another question that does not always result in such a resounding note of assurance, however. If we asked these same maintenance operators to explain to us in very basic terms how electricity works to make their plant equipment operate, the answers we probably would receive would be varied, jumbled, disjointed—and probably not all that accurate. Even on a more basic level, how many operators would be able to accurately answer the question: What is electricity?

Probably very few. Why do so many plant maintenance operators in both water and wastewater know so little about electricity? Part of the answer resides in the fact that maintenance operators are expected to know so much (and they are—and do) but are given so little opportunity to be properly trained.

We all know that experience is the great trainer. As an example, let's look at what a maintenance operator assigned to change the bearings on a 5-hp three-phase motor would need to know to accomplish this task. The operator would have to know:

1. How to deenergize the equipment (i.e., proper lockout/tagout procedures)
2. Once deenergized, how to properly disassemble the motor coupling from the device it operates (e.g., a motor coupling from a pump shaft) and the proper tools to use
3. Once uncoupled, how to properly disassemble the motor end-bells (preferably without damaging the rotor shaft)
4. Once disassembled, how to recognize if the bearings are really in need of replacement (though once removed from the end-bells, the bearings are typically replaced)

Questions the maintenance operator would need to know how to find answers for include the following:

1. If the bearings are in need of replacement, how are they to be removed without causing damage to the rotor shaft?
2. Once removed, what bearings should be used to replace the old bearings?
3. When the proper bearings are identified and obtained, how are they to be installed properly?
4. When the bearings are replaced properly, how is the motor to be reassembled properly?
5. Once the motor is correctly put back together, how is it properly aligned to the pump and then reconnected?
6. What is the test procedure to ensure that the motor has been restored properly to full operational status?

Every one of the steps and questions on the above procedures is important—obviously, errors at any point in the procedure could cause damage (maybe more damage than occurred in the first place). Here's another question to go along with that one—does the maintenance operator need to know electricity to perform the sequence of tasks described above?

The short answer is no, not exactly. Fully competent maintenance operators (who received most of their training via on-the-job experience) are usually qualified to perform the bearing-change-out activity on most plant motors with little difficulty.

The long answer is yes. Consider the motor mechanic who tunes your automobile engine. Then ask yourself: is it important that the mechanic have some understanding of internal combustion engines? We think it is important. You probably do, too. We also think it is important for the maintenance operator who changes bearings on an electrical motor to have some understanding of how the electric motor operates.

Here is another issue to look at. Have you ever taken a maintenance operator's state licensure examination? If you have, then you know that, typically, state licensure examinations test the examinee's knowledge of basic electricity. So, the states certainly consider operator knowledge of electricity important.

For reasons of licensure and of job competence, water/wastewater maintenance operators should have "some" basic electrical knowledge. But how and where can operators quickly and easily learn this important information?

In the *Fundamentals for the Water and Wastewater Maintenance Operator Series: Electricity,* we provide the how and the where.

1.2 NATURE OF ELECTRICITY

The word *electricity* is derived from the Greek word "electron" (meaning amber). Amber is a translucent (semitransparent) yellowish fossilized mineral resin. The ancient Greeks used the words "electric force" in referring to the mysterious forces of attraction and repulsion exhibited by amber when it was rubbed with a cloth. They did not understand the nature of this force. They could not answer the question, "What is electricity?" This question still remains unanswered. Today, we often attempt to answer this question by describing the effect and not the force. That is, the standard answer given is, "the force that moves electrons" is electricity, which is about the same as defining a sail as "the force that moves a sailboat."

At the present time, little more is known than the ancient Greeks knew about the fundamental nature of electricity, but we have made tremendous strides in harnessing and using it. As with many other unknown (or unexplainable) phenomena, elaborate theories concerning the nature and behavior of electricity have been advanced and have gained wide acceptance because of their apparent truth—and because they work.

Scientists have determined that electricity seems to behave in a constant and predictable manner in given situations or when subjected to given conditions. Faraday, Ohm, Lenz, and Kirchhoff have described the predictable characteristics of electricity and electric current in the form of certain rules. These rules are often referred to as laws. Thus, though electricity itself has never been clearly defined, its predictable nature and easily used form of energy has made it one of the most widely used power sources in modern times.

The bottom line: You can "learn" about electricity by learning the rules, or laws, applying to the behavior of electricity and by understanding the methods of producing, controlling, and using it. Thus, this learning can be accomplished without ever having determined its fundamental identity.

You are probably scratching your head, puzzled.

We understand the main question running through your brain at this exact moment: "This is a text about basic electricity, and the authors can't even explain what electricity is?"

That is correct; we cannot. The point is, no one can definitively define electricity. Electricity is one of those subject areas where the old saying "we don't know what we don't know about it" fits perfectly.

Again, there are a few theories about electricity that have so far stood the test of extensive analysis and much time (relatively speaking, of course). One of the oldest and most generally accepted theories concerning electric current flow (or electricity) is known as the *electron theory.*

The electron theory basically states that electricity or current flow is the result of the flow of free electrons in a conductor. Thus, electricity is the flow of free electrons or simply electron flow. And, in this text, this is how we define electricity; that is, again, electricity is the flow of free electrons.

Electrons are extremely tiny particles of matter. To gain understanding of electrons and exactly what is meant by "electron flow," it is necessary to briefly discuss the structure of matter.

1.3 THE STRUCTURE OF MATTER

Matter is anything that has mass and occupies space. To study the fundamental structure or composition of any type of matter, it must be reduced to its fundamental components. All matter is made of *molecules,* or combinations of *atoms* (Greek: not able to be divided), that are bound together to produce a given substance, such as salt, glass, or water. For example, if you keep dividing water into smaller and smaller drops, you would eventually arrive at the smallest particle that was still water. That particle is the molecule, which is defined as the **smallest bit of a substance that retains the characteristics of that substance.**

A molecule of water (H_2O) is composed of one atom of oxygen and two atoms of hydrogen, as represented in Figure 1.1. (The three molecules are not in a straight line; instead, they form an angle of 105°.) If the molecule of water was further subdivided, there would remain only unrelated atoms of oxygen and hydrogen, and the water would no longer exist as such. Thus, the molecule is the smallest particle to which a substance can be reduced and still be called by the same name. This applies to all substances—solids, liquids, and gases.

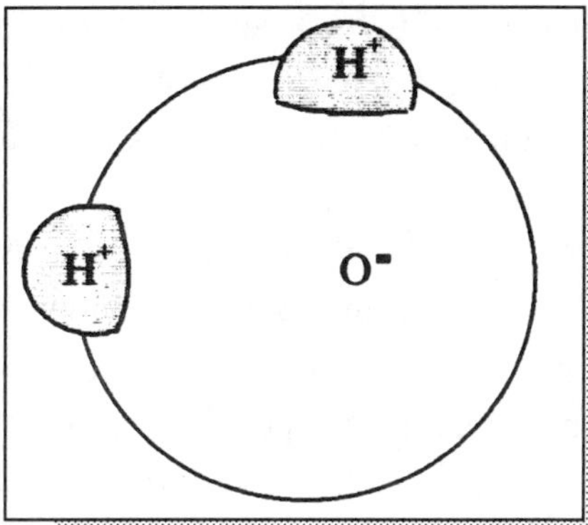

Figure 1.1
A molecule of water.

IMPORTANT

Important Point: *Molecules are made up of atoms, which are bound together to produce a given substance.*

Atoms are composed, in various combinations, of subatomic particles of *electrons, protons, and neutrons.* These particles differ in weight (a proton is much heavier than an electron) and charge. We are not concerned with the weights of particles in this text, but the **charge** is extremely important in electricity. The electron is the fundamental nega-

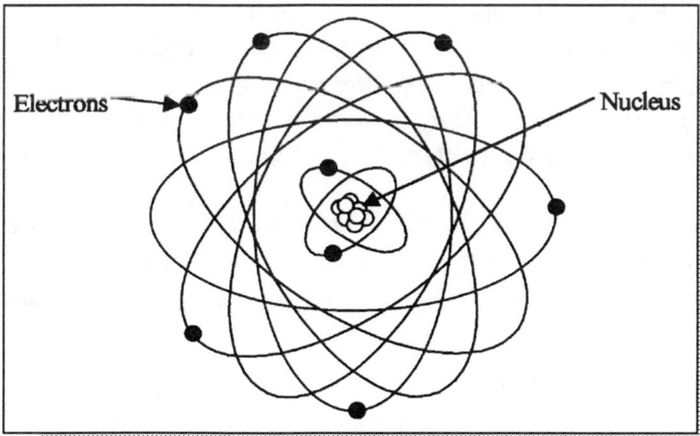

Figure 1.2
Electrons and nucleus of an atom.

tive charge (–) of electricity. Electrons revolve about the nucleus or center of the atom in paths of concentric *orbits,* or shells (see Figure 1.2). The proton is the fundamental positive (+) charge of electricity. Protons are found in the nucleus. The number of protons within the nucleus of any particular atom specifies the atomic number of that atom. For example, the helium atom has two protons in its nucleus so the atomic number is 2. The neutron, which is the fundamental neutral charge of electricity, is also found in the nucleus.

Most of the weight of the atom is in the protons and neutrons of the nucleus. Whirling around the nucleus are one or more negatively charged electrons. Normally, there is one proton for each electron in the entire atom so that the net positive charge of the nucleus is balanced by the net negative charge of the electrons rotating around the nucleus (see Figure 1.3).

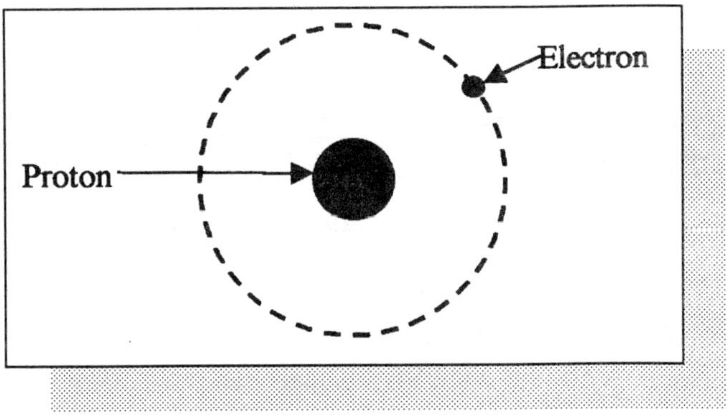

Figure 1.3
One proton and one electron = electrically neutral.

IMPORTANT

Important Point: *Most batteries are marked with the symbols + and – or even with the abbreviations POS (positive) and NEG (negative). The concept of a positive or negative polarity and its importance in electricity will become clear later. However, for the moment, you need to remember that an electron has a negative charge and that a proton has a positive charge.*

We stated earlier that in an atom the number of protons is usually the same as the number of electrons. This is an important point because this relationship determines the kind of element (the atom is the smallest particle that makes up an element; an element retains its characteristics when subdivided into atoms) in question. Figure 1.4 shows a simplified drawing of several atoms of different materials based on the conception of electrons orbiting about the nucleus. For example, hydrogen has a nucleus consisting of one proton, around which rotates one electron. The helium atom has a nucleus containing two protons and two neutrons with two electrons encircling the nucleus. Both of these elements are electrically neutral (or balanced) because each has an equal number of electrons and protons. Because the negative (–) charge of each electron is equal in magnitude to the positive (+) charge of each proton, the two opposite charges cancel.

A balanced (neutral or stable) atom has a certain amount of energy, which is equal to the sum of the energies of its electrons. Electrons, in turn, have different energies called *energy levels*. The energy level of an electron is proportional to its distance from the nucleus. Therefore, the energy levels of electrons in shells farther from the nucleus are higher than those of electrons in shells nearer the nucleus.

When an electric force is applied to a conducting medium, such as copper wire, electrons in the outer orbits of the copper atoms are forced out of orbit (i.e., liberating or freeing electrons) and are impelled along the wire. This electrical force, which forces electrons out of orbit, can be

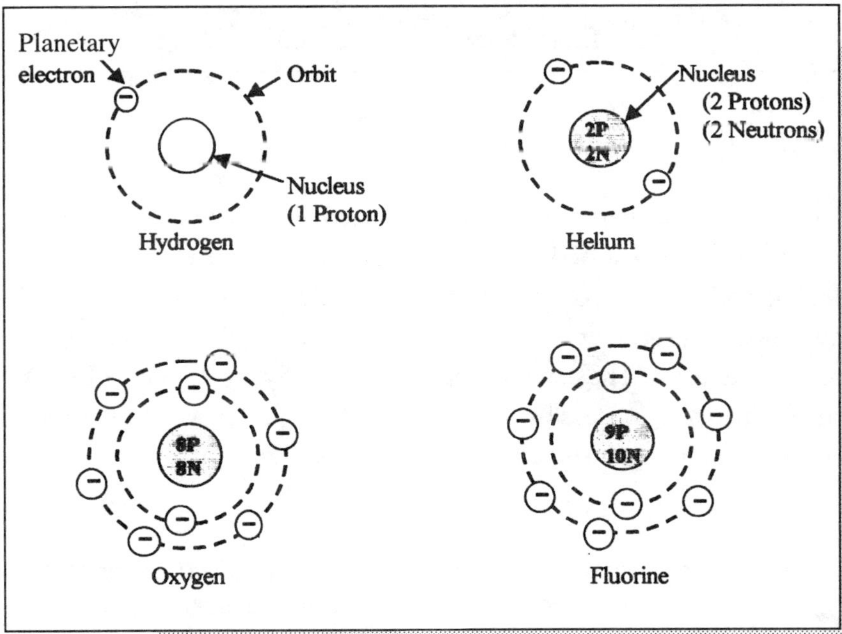

Figure 1.4
Atomic structure of elements.

produced in a number of ways, such as by moving a conductor through a magnetic field; by friction, as when a glass rod is rubbed with cloth (silk); or by chemical action, as in a battery. [The principal methods of producing this force (voltage) are described in Section 1.7.2].

When the electrons are forced from their orbits, they are called **free electrons.** Some of the electrons of certain metallic atoms are so loosely bound to the nucleus that they are relatively free to move from atom to atom. These free electrons constitute the flow of an electric current in electrical conductors.

IMPORTANT ***Important Point:*** *When an electric force is applied to a copper wire, free electrons are displaced from the copper atoms and move along the wire, producing electric current as shown in Figure 1.5.*

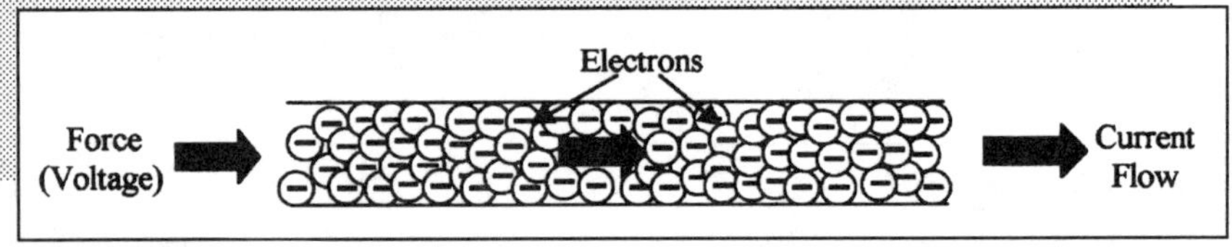

Figure 1.5
Electron flow in a copper wire.

If the internal energy of an atom is raised above its normal state, the atom is said to be **excited.** Excitation may be produced by causing the atoms to collide with particles that are impelled by an electric force as shown in Figure 1.5. In effect, energy is transferred from the electric source to the atom. The excess energy absorbed by an atom may become sufficient to cause loosely bound outer electrons (as shown in Figure 1.5) to leave the atom against the force that acts to hold them within.

IMPORTANT ***Important Point:*** *An atom that has lost or gained one or more electrons is said to be* ***ionized.*** *If the atom loses electrons, it becomes positively charged and is referred to as a* ***positive ion.*** *Conversely, if the atom gains electrons, it becomes negatively charged and is referred to as a* ***negative ion.***

1.4 CONDUCTORS, SEMICONDUCTORS, AND INSULATORS

Electric current moves easily through some materials but with greater difficulty through others. Substances that permit the free movement of a large number of electrons are called *conductors.* The most widely used electrical conductor is copper because of its high conductivity (how good a conductor the material is) and cost-effectiveness.

Electrical energy is transferred through a copper or other metal conductor by means of the movement of free electrons that migrate from atom to atom inside the conductor (see Figure 1.5). Each electron moves a very short distance to the neighboring atom where it replaces one or more electrons by forcing them out of their orbits. The replaced electrons repeat the process in other nearby atoms until the movement is transmitted throughout the entire length of the conductor. A good conductor is said to have a low opposition, or **resistance,** to the electron (current) flow.

IMPORTANT ***Note:*** *If lots of electrons flow through a material with only a small force (voltage) applied, we call that material a* ***conductor.***

TABLE 1.1. Electrical Conductors.
Silver
Copper
Gold
Aluminum
Zinc
Brass
Iron
Tin
Mercury

Table 1.1 lists many of the metals commonly used as electrical conductors. The best conductors appear at the top of the list, with the poorer ones shown last.

IMPORTANT ***Note:*** *The movement of each electron (e.g., in copper wire) takes a very small amount of time. It happens almost instantly. This is an important point to keep in mind later in the book, when events in an electrical circuit seem to occur simultaneously.*

While it is true that electron motion is known to exist to some extent in all matter, some substances, such as rubber, glass, and dry wood, have very few free electrons. In these materials, large amounts of energy must be expended in order to break the electrons loose from the influence of the nucleus. Substances containing very few free electrons are called *insulators*. Insulators are important in electrical work because they prevent the current from being diverted from the wires.

IMPORTANT

***Note:** If the voltage is large enough, even the best insulators will break down and allow their electrons to flow.*

Table 1.2 lists some materials that we often use as insulators in electrical circuits. The list is in decreasing order of ability to withstand high voltages without conducting.

TABLE 1.2. Common Insulators.

Rubber
Mica
Wax or paraffin
Porcelain
Bakelite
Plastics
Glass
Fiberglass
Dry wood
Air

A material that is neither a good conductor nor a good insulator is called a *semiconductor*. Silicon and germanium are substances that fall into this category. Because of their peculiar crystalline structure, these materials may act as conductors under certain conditions; under other conditions, they act as insulators. As the temperature is raised, however, a limited number of electrons becomes available for conduction.

1.5 STATIC ELECTRICITY

Electricity at rest is often referred to as *static electricity*. More specifically, when two bodies of matter have unequal charges, and are near one another, an electric force is exerted between them because of their unequal charges. However, because they are not in contact, their charges cannot equalize. The existence of such an electric force, where current can't flow, is **static electricity.**

However, static, or electricity at rest, will flow if given the opportunity. An example of this phenomenon is often experienced when one walks across a dry carpet and then touches a doorknob—a slight shock is usually felt, and a spark at the fingertips is likely noticed. In the workplace, static electricity is prevented from building up by properly bonding equipment to ground or earth.

1.5.1 CHARGED BODIES

To fully grasp a clear understanding of static electricity, it is necessary to know one of the fundamental laws of electricity and its significance.

The fundamental law of charged bodies states, "Like charges repel each other and unlike charges attract each other."

A positive charge and a negative charge, being opposite or unlike, tend to move toward each other—attracting each other. In contrast, like bodies tend to repel each other. Electrons repel each other because of their like negative charges, and protons repel each other because of their like positive charges. Figure 1.6 demonstrates the law of charged bodies.

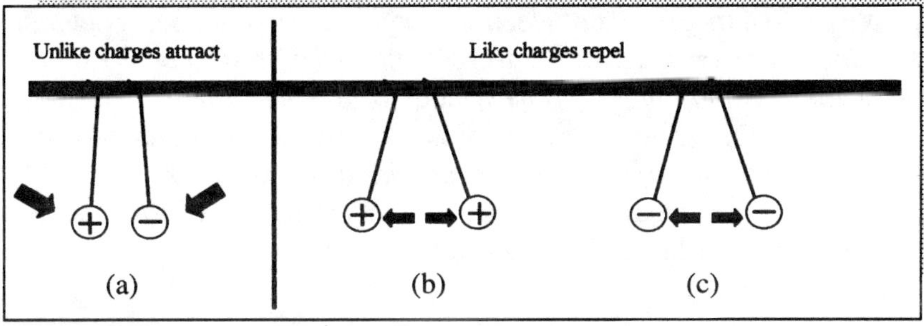

Figure 1.6
Reaction between two charged bodies. The opposite charge in (a) attracts. The like charges in (b) and (c) repel each other.

It is important to point out another significant part of the fundamental law of charged bodies—that is, *force of attraction or repulsion existing between two magnetic poles decreases rapidly as the poles are separated from each other.* More specifically, the force of attraction or repulsion varies directly as the product of the separate pole strengths and inversely as the square of the distance separating the magnetic poles, provided the poles are small enough to be considered as points.

Let's look at an example.

If you increase the distance between two north poles from 2 feet to 4 feet, the force of repulsion between them is decreased to one fourth of its original value. If either pole strength is doubled, the distance remaining the same, the force between the poles will be doubled.

1.5.2 COULOMB'S LAW

Simply put, Coulomb's Law points out that the amount of attracting or repelling force that acts between two electrically charged bodies in free space depends on two things:

a. Their charges

b. The distance between them.

Specifically, Coulomb's Law states, **"Charged bodies attract or repel each other with a force that is directly proportional to the product of their charges and is inversely proportional to the square of the distance between them."**

IMPORTANT *__Note:__ The magnitude of electric charge a body possesses is determined by the number of electrons compared with the number of protons within the body. The symbol for the magnitude of electric charge is Q, expressed in units of __coulombs (C)__. A charge of one positive coulomb means a body contains a charge of 6.25×10^{18}. A charge of one negative coulomb, −Q, means a body contains a charge of 6.25 times 10^{18} more electrons than protons.*

1.5.3 ELECTROSTATIC FIELDS

The fundamental characteristic of an electric charge is its ability to exert force. The space between and around charged bodies in which their influence is felt is called an *electric field of force.* The electric field is always terminated on material objects and extends between positive and negative charges. This region of force can consist of air, glass, paper, or a vacuum. This region of force is referred to as an *electrostatic field.*

When two objects of opposite polarity are brought near each other, the electrostatic field is concentrated in the area between them. The field is generally represented by lines that are referred to as **electrostatic lines of force.** These lines are imaginary and are used merely to represent the direction and strength of the field. To avoid confusion, the positive lines of force are always shown leaving charge, and, for a negative charge, they are shown as entering. Figure 1.7 illustrates the use of lines to represent the field about charged bodies.

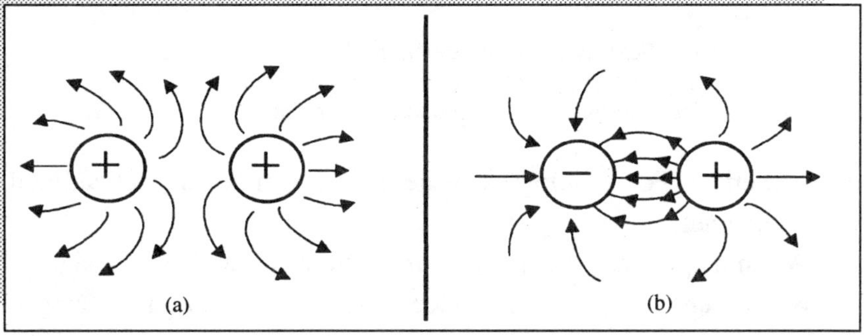

Figure 1.7
Electrostatic lines of force. (a) Represents the repulsion of like-charged bodies and their associated fields. (b) Represents the attraction between unlike-charged bodies and their associated fields.

IMPORTANT **Note:** *A charged object will retain its charge temporarily if there is no immediate transfer of electrons to or from it. In this condition, the charge is said to be* at rest. *Remember, electricity at rest is called* static *electricity.*

1.6 MAGNETISM

Most electrical equipment depends directly or indirectly upon magnetism. Magnetism is defined as a phenomenon associated with magnetic fields; that is, it has the power to attract such substances as iron, steel, nickel, or cobalt (metals that are known as magnetic materials). Correspondingly, a substance is said to be a magnet if it has the property

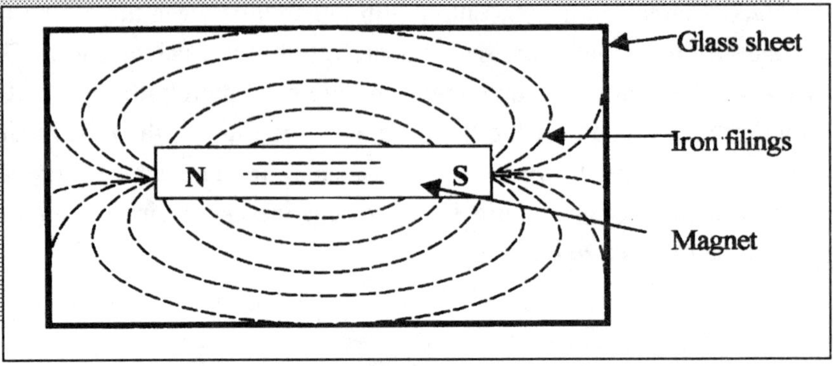

Figure 1.8
The magnetic field around a bar magnet. If the glass sheet is tapped gently, the filings will move into a definite pattern that describes the field of force around the magnet.

of magnetism. For example, a piece of iron can be magnetized and is, thus, a magnet.

When magnetized, the piece of iron (Note: we will assume a piece of flat bar 6 inches long x 1 inch wide x 0.5 inches thick; a bar magnet—see Figure 1.8) will have two points opposite each other that most readily attract other pieces of iron. The points of maximum attraction (one on each end) are called the *magnetic poles* of the magnet: the north (N) pole and the south (S) pole. Just as like electric charges repel each other and opposite charges attract each other, like magnetic poles repel each other, and unlike poles attract each other. Although invisible to the naked eye, its force can be shown to exist by sprinkling small iron filings on a glass covering a bar magnet as shown in Figure 1.8.

Figure 1.9
Magnetic field of force around a bar magnet, indicated by lines of force.

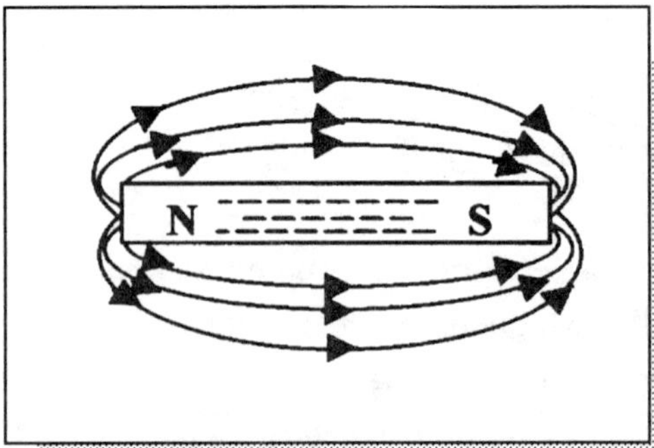

Figure 1.9 shows how the field looks without iron filings; it is shown as lines of force [known as **magnetic flux** or **flux lines;** the symbol for magnetic flux is the Greek lowercase letter ϕ (phi)] in the field, repelled away from the north pole of the magnet and attracted to its south pole.

IMPORTANT

***Note:** A **magnetic circuit** is a complete path through which magnetic lines of force may be established under the influence of a magnetizing force. Most magnetic circuits are composed largely of magnetic materials in order to contain the magnetic flux. These circuits are similar to the **electric circuit** (an important point), which is a complete path through which current is caused to flow under the influence of an electromotive force.*

There are three types or groups of magnets:

a. *Natural Magnets:* found in the natural state in the form of a mineral (an iron compound) called magnetite.
b. *Permanent Magnets:* artificial magnet, hardened steel or some alloy such as Alinco bars that have been permanently magnetized. The permanent magnet with which most people are familiar is the horseshoe magnet (see Figure 1.10).
c. *Electromagnets:* artificial magnet, composed of soft-iron cores around which are wound coils of insulated wire. When an electric current flows through the coil, the core becomes magnetized. When the current ceases to flow, the core loses most of the magnetism. (Note: Electromagnets and electromagnetism are discussed in greater detail in Chapter 5—Electromagnetism).

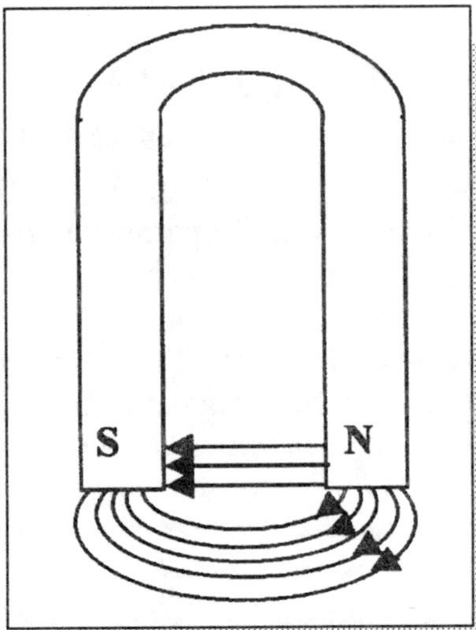

Figure 1.10
Horseshoe magnet.

1.6.1 MAGNETIC MATERIALS

Natural magnets are no longer used (they have no practical value) in electrical circuitry because more powerful and more conveniently shaped permanent magnets can be produced artificially. Commercial magnets are made from special steels and alloys—magnetic materials.

Magnetic materials are those materials that are attracted or repelled by a magnet and that can be magnetized themselves. Iron, steel, and alloy bar are the most common magnetic materials. These materials can be magnetized by inserting the material (in bar form) into a coil of insulated wire and passing a heavy direct current through the coil. (This aspect of magnetism is treated later in Chapter 5). The same material may also be magnetized if it is stroked with a bar magnet. It will then have the same magnetic property that the magnet used to induce the magnetism has—namely, there will be two poles of attraction, one at either end. This process produces a permanent magnet by **induction**—that is, the magnetism is induced in the bar by the influence of the stroking magnet.

IMPORTANT ***Note:*** *Permanent magnets are those of hard magnetic materials (hard steel or alloys) that retain their magnetism when the magnetizing field is removed. A temporary magnet is one that has no ability to retain a magnetized state when the magnetizing field is removed.*

Even though classified as permanent magnets, it is important to point out that hardened steel and certain alloys are relatively difficult to magnetize and are said to have a **low permeability** because the magnetic lines of force do not easily permeate or distribute themselves readily through the steel.

IMPORTANT ***Note: Permeability*** *refers to the ability of a magnetic material to concentrate magnetic flux. Any material that is easily magnetized has high permeability. A measure of permeability for different materials in comparison with air or vacuum is called* ***relative*** *permeability, symbolized by m or (mu).*

Once hard steel and other alloys are magnetized, however, they retain a large part of their magnetic strength and are called **permanent mag-**

nets. Conversely, materials that are relatively easy to magnetize—such as soft iron and annealed silicon steel—are said to have a **high permeability.** Such materials retain only a small part of their magnetism after the magnetizing force is removed and are called **temporary magnets.**

The magnetism that remains in a temporary magnet after the magnetizing force is removed is called **residual magnetism.** Residual magnetism is important and is discussed in greater detail in Section 11.2.1 (D-C Generators).

Early magnetic studies classified magnetic materials merely as being magnetic and nonmagnetic—that is, based on the strong magnetic properties of iron. However, because weak magnetic materials can be important in some applications, present studies classify materials into one of three groups: paramagnetic, diamagnetic, and ferromagnetic.

a. *Paramagnetic materials:* These include aluminum, platinum, manganese, and chromium—materials that become only slightly magnetized even though under the influence of a strong magnetic field. This slight magnetization is in the same direction as the magnetizing field. Relative permeability is slightly more than 1 (i.e., considered nonmagnetic materials).
b. *Diamagnetic materials:* These include bismuth, antimony, copper, zinc, mercury, gold, and silver—materials that can also be slightly magnetized when under the influence of a very strong field. Relative permeability is less than 1 (i.e., considered nonmagnetic materials).
c. *Ferromagnetic materials:* These include iron, steel, nickel, cobalt, and commercial alloys—materials that are the most important group for applications of electricity and electronics. Ferromagnetic materials are easy to magnetize and have high permeability, ranging from 50 to 3000.

1.6.2 MAGNETIC EARTH

The earth is a huge magnet; and surrounding earth is the magnetic field produced by the earth's magnetism. Most people would have no problem understanding or at least accepting this statement. However, if told that the earth's north magnetic pole is actually its south magnetic pole and that the south magnetic pole is actually the earth's north magnetic pole, they might not accept or understand this statement. But, in terms of a magnet, it is true.

As can be seen from Figure 1.11, the magnetic polarities of the earth are indicated. The geographic poles are also shown at each end of the axis of rotation of the earth. Clearly, as shown in Figure 1.11, the magnetic axis does not coincide with the geographic axis, and, therefore, the magnetic and geographic poles are not at the same place on the surface of the earth.

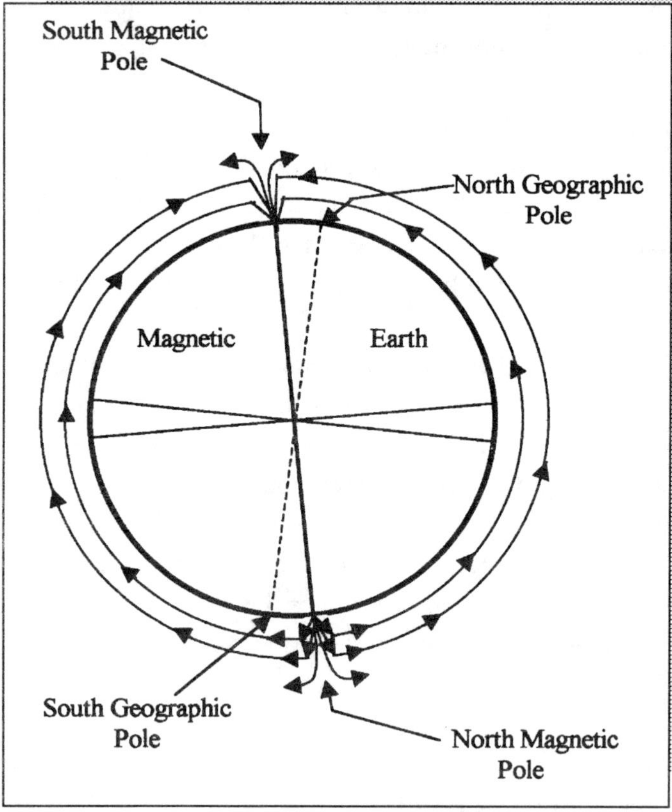

Figure 1.11
Earth's magnetic poles.

Recall that magnetic lines of force are assumed to emanate from the north pole of a magnet and to enter the south pole as closed loops. Because the earth is a magnet, lines of force emanate from its north magnetic pole and enter the south magnetic pole as closed loops. A compass needle aligns itself in such a way that the earth's lines of force enter at its south pole and leave at its north pole. Because the north pole of the needle is defined as the end that points in a northerly direction, it follows that the magnetic pole in the vicinity of the north geographic pole is in reality a south magnetic pole, and vice versa.

1.7 DIFFERENCE IN POTENTIAL

Because of the force of its electrostatic field, an electric charge has the ability to do the work of moving another charge by attraction or repulsion. The force that causes free electrons to move in a conductor as an electric current may be referred to as follows:

- electromotive force (emf)
- voltage
- difference in potential

When a difference in potential exists between two charged bodies that are connected by a wire (conductor), electrons (current) will flow along the conductor. This flow is from the negatively charged body to the positively charged body until the two charges are equalized and the potential difference no longer exists.

IMPORTANT **Note:** *The basic unit of potential difference is the* volt *(V). The symbol for potential difference is* V, *indicating the ability to do the work of forcing electrons (current flow) to move. Because the volt unit is used, potential difference is called* voltage.

1.7.1 THE WATER ANALOGY

In attempting to train individuals in the concepts of basic electricity, especially in regard to difference of potential (voltage), current and resistance relationships in a simple electrical circuit, it has been common practice to use what is referred to as the water analogy. We use the water analogy later to explain (in simple straightforward fashion) voltage, current, and resistance and their relationships in more detail, but for now we use the analogy to explain the basic concept of electricity: difference of potential, or voltage. Because a difference in potential causes current flow (against resistance), it is important that this concept be understood first before the concept of current flow and resistance are explained.

Consider the water tanks shown in Figure 1.12—two water tanks connected by a pipe and valve. At first, the valve is closed, and all the water is in Tank A. Thus, the water pressure across the valve is at maximum. When the valve is opened, the water flows through the pipe from A to B until the water level becomes the same in both tanks. The water then stops flowing in the pipe, because there is no longer a difference in water pressure (difference in potential) between the two tanks.

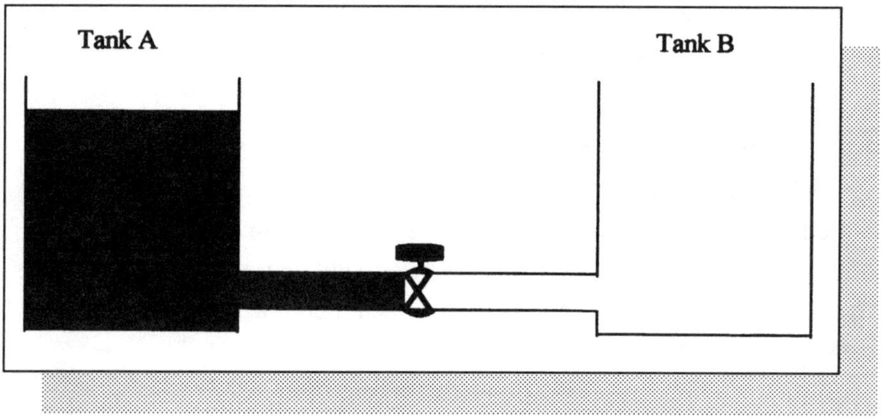

Figure 1.12
Water analogy of electric difference of potential.

Just as the flow of water through the pipe in Figure 1.12 is directly proportional to the difference in water level in the two tanks, current flow through an electric circuit is directly proportional to the difference in potential across the circuit.

IMPORTANT ***Important Point:*** *A fundamental law of current electricity is that the **current is directly proportional to the applied voltage;** that is, if the voltage is increased, the current is increased. If the voltage is decreased, the current is decreased.*

1.7.2 PRINCIPAL METHODS OF PRODUCING A VOLTAGE

There are many ways to produce electromotive force, or voltage. Some of these methods are much more widely used than others. The following is a list of the six most common methods of producing electromotive force.

a. **Friction:** voltage produced by rubbing two materials together.

b. **Pressure (Piezoelectricity):** voltage produced by squeezing crystals of certain substances.

c. **Heat (Thermoelectricity):** voltage produced by heating the joint (junction) where two unlike metals are joined.

d. **Light (Photoelectricity):** voltage produced by light striking photosensitive (light sensitive) substances.

e. **Chemical Action:** voltage produced by chemical reaction in a battery cell.

f. **Magnetism:** voltage produced in a conductor when the conductor moves through a magnetic field or a magnetic field moves through the conductor in such a manner as to cut the magnetic lines of force of the field.

In the study of basic electricity, we are most concerned with magnetism (generators; see Chapter 11) and chemistry (batteries; see Chapter 2) as means to produce voltage. Friction has few practical applications, though we discussed it earlier in studying static electricity. Pressure, heat, and light do have useful applications, but we do not need to consider them in this text. Magnetism and chemistry, on the other hand, are the principal sources of voltage and are discussed at length in this text.

1.8 ELECTRIC CURRENT

The movement or the flow of electrons is called *current.* To produce current, the electrons must be moved by a potential difference.

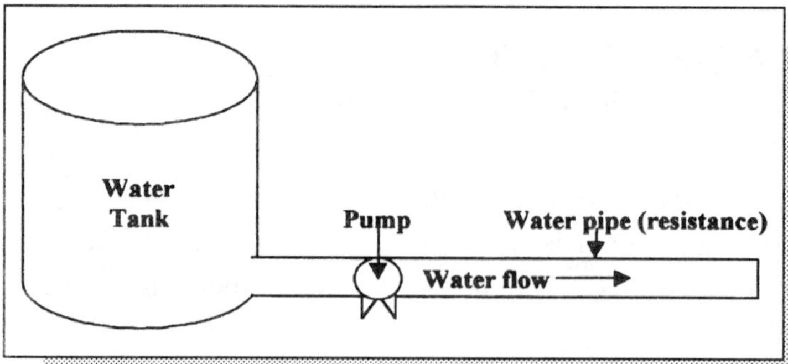

Figure 1.13
Water analogy: current flow.

IMPORTANT **Note:** *The terms current, current flow, electron flow, or electron current, etc., may be used to describe the same phenomenon.*

Electron flow, or current, in an electric circuit is from a region of less negative potential to a region of more positive potential.

IMPORTANT **Note:** *Current is represented by the letter I. The basic unit in which current is measured is the* ***ampere****, or* ***amp (A)****. One ampere of current is defined as the movement of one coulomb past any point of a conductor during one second of time.*

Recall that in Section 1.6 we used the water analogy to help us understand potential difference. We can also use the water analogy to help us understand current flow through a simple electric circuit.

Consider Figure 1.13 that shows a water tank connected via a pipe to a pump with a discharge pipe. If the water tank contains an amount of water above the level of the pipe opening to the pump, the water exerts pressure (a difference in potential) against the pump. When sufficient water is available for pumping with the pump, water flows through the

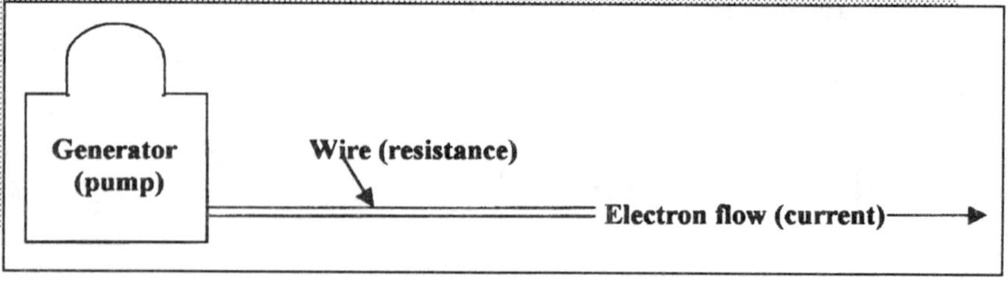

Figure 1.14
Simple electric circuit with current flow.

pipe against the resistance of the pump and pipe. The analogy should be clear—in an electric circuit, if a difference of potential exists, current will flow in the circuit.

Another simple way of looking at this analogy is to consider Figure 1.14 where the water tank has been replaced with a generator, the pipe with a conductor (wire), and water flow with the flow of electric current.

Again, the key point illustrated by Figures 1.13 and 1.14 is that, to produce current, the electrons must be moved by a potential difference.

Electric current is generally classified into two general types:

- direct current (d-c)
- alternating current (a-c)

Direct current is current that moves through a conductor or circuit in one direction only. **Alternating current** periodically reverses direction. These two types of current are discussed in greater detail later in Chapters 3 and 6.

1.9 RESISTANCE

In Section 1.3, we discussed conductors and insulators. We pointed out that free electrons, or electric current, could move easily through a good conductor, such as copper, but that an insulator, such as glass, was an obstacle to current flow. In the water analogy shown in Figure 1.13 and the simple electric circuit shown in Figure 1.14, resistance is indicated by either the pipe or the conductor.

Every material offers some resistance, or opposition, to the flow of electric current through it. Good conductors, such as copper, silver, and aluminum, offer very little resistance. Poor conductors, or insulators, such as glass, wood, and paper, offer a high resistance to current flow.

IMPORTANT

Note: *The amount of current that flows in a given circuit depends on two factors: voltage and resistance.*

IMPORTANT

Note: *Resistance is represented by the letter* R. *The basic unit in which resistance is measured is the* ***ohm*** *(Ω). One ohm is the resistance of a circuit element, or circuit, that permits a steady current* *of 1 ampere (1 coulomb per second) to flow when a steady electromotive force (emf) of 1 volt is applied to the circuit. Manufactured circuit parts containing definite amounts of resistance are called* ***resistors.***

The size and type of material of the wires in an electric circuit are chosen so as to keep the electrical resistance as low as possible. In this way, current can flow easily through the conductors, just as water flows through the pipe between the tanks in Figure 1.12. If the water pressure remains constant, the flow of water in the pipe will depend on how far the valve is opened. The smaller the opening, the greater the opposition (resistance) to the flow, and the smaller will be the rate of flow in gallons per second.

In the electric circuit shown in Figure 1.14, the larger the diameter of the wire, the lower will be its electrical resistance (opposition) to the flow of current through it. In the water analogy, pipe friction opposes the flow of water between the tanks. This friction is similar to electrical resistance. The resistance of the pipe to the flow of water through it depends upon (1) the length of the pipe, (2) diameter of the pipe, and (3) the nature of the inside walls (rough or smooth). Similarly, the electrical resistance of the conductors depends upon (1) the length of the wires, (2) the diameter of the wires, and (3) the material of the wires (copper, silver, etc.).

It is important to note that temperature also affects the resistance of electrical conductors to some extent. In most conductors (copper, aluminum, etc.) the resistance increases with temperature. Carbon is an exception. In carbon, the resistance decreases as temperature increases.

IMPORTANT

Important Note: *Electricity is a study that is frequently explained in terms of opposites. The term that is exactly the opposite of resistance is* ***conductance.*** *Conductance (G) is the ability of a material to pass electrons. The unit of conductance is the* ***Mho,*** *which is ohm spelled backwards. The relationship that exists between resistance and conductance is the reciprocal. A reciprocal of a number is obtained by dividing the number into one. If the resistance of a material is known, dividing its value into one will give its conductance. Similarly, if the conductance is known, dividing its value into one will give its resistance.*

Self-Test

The answers to chapter self-tests are found in Appendix A.

1.1 An electrical circuit with a conductance of 5 mho would have a resistance of ____________.

1.2 Electrons move about the nucleus of an atom in paths that are called ____________.

1.3 The nucleus of an atom consists of particles called ____________ and ____________.

1.4 The number of protons in the nucleus of an atom is known as the ____________ ____________ of an atom.

1.5 Unlike charges ____________ each other, while like charges ____________ each other.

1.6 Define in your own words the expression "difference in potential."

1.7 List the six ways in which voltage is produced.

1.8 Give the unit of measurement and symbol for each of the following: (a) difference in potential; (b) rate of current flow; (c) quantity of electricity; (d) resistance.

1.9 What three factors affect the resistance in a circuit?

1.10 What is the difference between direct and alternating current?

1.11 What are the points of maximum attraction on a magnet?

1.12 What is a magnetic field?

1.13 What are the three general groups of magnets?

1.14 What are the two most widely used methods of producing a voltage?

1.15 The rate of current flow in a circuit can be changed by changing what other two variables?

Battery-Supplied Electricity

TOPICS

The Voltaic Cell
Battery
Types of Batteries
Battery Characteristics

2.1 INTRODUCTION

Battery-supplied direct current electricity has many applications and is widely used in water/wastewater treatment operations. Applications include providing electrical energy in plant vehicles and emergency diesel generators, material handling equipment (forklifts), and portable electric/electronic equipment; and providing backup emergency power for light-packs and for hazard warning signal lights and flashlights; and as standby power supplies or uninterruptable power supplies (UPS) for computer systems. In some instances, they are used as the only source of power; while in others (as mentioned above), they are used as a secondary or standby power supply.

2.2 THE VOLTAIC CELL

The simplest cell (a device that transforms chemical energy into electrical energy) is known as a *voltaic* (or galvanic) cell (see Figure 2.1). It consists of a piece of carbon (C) and a piece of zinc (Zn) suspended in a jar that contains a solution of water (H_2O) and sulfuric acid (H_2SO_4).

IMPORTANT ***Note:*** *A simple cell consists of two strips, or* ***electrodes****, placed in a container that holds the* ***electrolyte****. A battery is formed when two or more cells are connected.*

Key Terms Used in This Chapter

BATTERY ▶ A device for producing d-c electric current from a chemical reaction.

CELL ▶ A device that transforms chemical energy into electrical energy.

ELECTRODES ▶ The conductors by which the current leaves or returns to the electrolyte.

ELECTROLYTE ▶ The solution that acts upon the electrodes that are placed in it.

SPECIFIC GRAVITY ▶ The ratio of the density of a substance to a standard density. For liquids, the density is compared with the density of water.

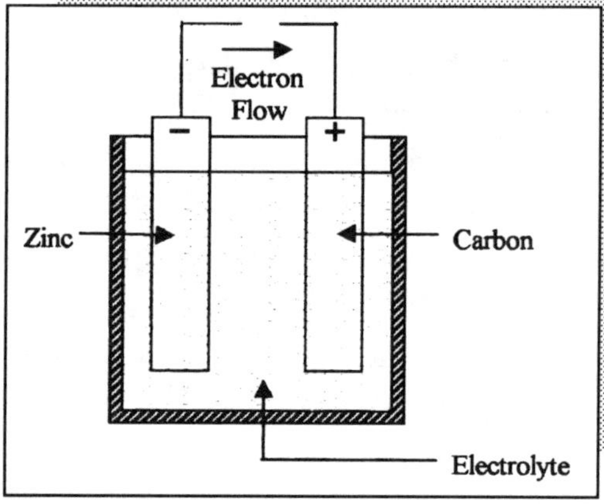

Figure 2.1 *Simple voltaic cell.*

The electrodes are the conductors by which the current leaves or returns to the electrolyte. In the simple cell described above, they are carbon and zinc strips placed in the electrolyte. Zinc contains an abundance of negatively charged atoms, while carbon has an abundance of positively charge atoms. When the plates of these materials are immersed in an electrolyte, chemical action between the two begins.

In the **dry cell** (see Figure 2.2), the electrodes are the carbon rod in the center and the zinc container in which the cell is assembled.

The electrolyte is the solution that acts upon the electrodes that are placed in it. The electrolyte may be a salt, an acid, or an alkaline solution. In the simple voltaic cell and in the automobile storage battery, the electrolyte is in a liquid form; while in the dry cell (see Figure 2.2), the electrolyte is a moist paste.

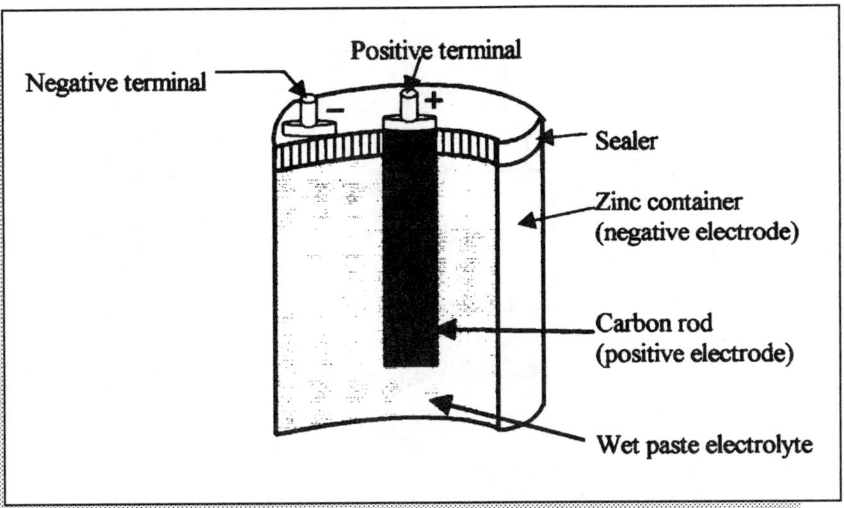

Figure 2.2
Dry cell (cross-sectional view).

2.2.1 PRIMARY AND SECONDARY CELLS

Primary cells are normally those that cannot be recharged or returned to good condition after their voltage drops too low. Dry cells in flashlights and transistor radios are examples of primary cells. Some primary cells have been developed to the state where they can be recharged.

A *secondary cell* is one in which the electrodes and the electrolyte are altered by the chemical action that takes place when the cell delivers current. These cells are rechargeable. During recharging, the chemicals that provide electric energy are restored to their original condition. Recharging is accomplished by forcing an electric current through them in the opposite direction to that of discharge.

A cell is recharged by connecting as shown in Figure 2.3. Some battery chargers have a voltmeter and an ammeter that indicate the charging voltage and current.

The automobile storage battery is the most common example of the secondary cell.

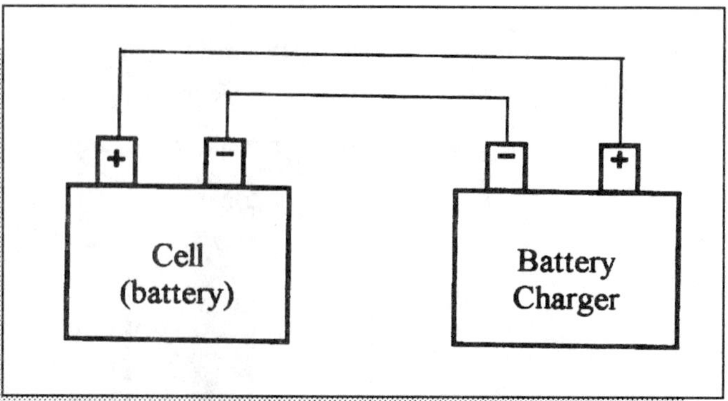

Figure 2.3 *Hookup for charging a secondary cell with a battery charger.*

2.3 BATTERY

As was stated previously, a battery consists of two or more cells placed in a common container. The cells are connected in series, in parallel, or in some combination of series and parallel, depending upon the amount of voltage and current required of the battery. The connection of cells in a battery is discussed in more detail in Section 2.3.2.

2.3.1 BATTERY OPERATION

The chemical reaction within a battery provides the voltage. This occurs when a conductor is connected externally to the electrodes of a cell, causing electrons to flow under the influence of a difference in potential across the electrodes from the zinc (negative) through the external conductor to the carbon (positive), returning within the solution to the zinc. After a short period of time, the zinc will begin to waste away because of the acid.

The voltage across the electrodes depends upon the materials from which the electrodes are made and the composition of the solution. The difference of potential between the carbon and zinc electrodes in a dilute solution of sulfuric acid and water is about 1.5 volts.

The current that a primary cell may deliver depends upon the resistance of the entire circuit, including that of the cell itself. The internal resistance of the primary cell depends upon the size of the electrodes, the distance between them in the solution, and the resistance of the solution. The larger the electrodes and the closer together they are in solution (without touching), the lower the internal resistance of the primary cell and the more current it is capable of supplying to the load.

IMPORTANT

Note: *When current flows through a cell, the zinc gradually dissolves in the solution, and the acid is neutralized.*

2.3.2 COMBINING CELLS

In many operations, battery-powered devices may require more electrical energy than one cell can provide. Various devices may require either a higher voltage or more current and, in some cases, both. Under such conditions, it is necessary to combine, or interconnect, a sufficient number of cells to meet the higher requirements. Cells connected in series provide a higher voltage, while cells connected in parallel provide a higher current capacity. To provide adequate power when both voltage and current requirements are greater than the capacity of one cell, a combination series-parallel network of cells must be interconnected.

When cells are connected in ***series*** (see Figure 2.4), the total voltage across the battery of cells is equal to the sum of the voltage of each of the individual cells. In Figure 2.4, the four 1.5-V cells in series provide a total battery voltage of 6 V. When cells are placed in series, the positive terminal of one cell is connected to the negative terminal of the other cell. The positive electrode of the first cell and negative electrode of the last cell then serve as the power takeoff terminals of the battery. The current flowing through such a battery of series cells is the same as from one cell because the same current flows through all the series cells.

To obtain a greater current, a battery has cells connected in **parallel** as shown in Figure 2.5. In this parallel connection, all the positive electrodes are connected to one line, and all negative electrodes are connected to the other. Any point on the positive side can serve as the positive terminal of the battery, and any point on the negative side can be the negative terminal.

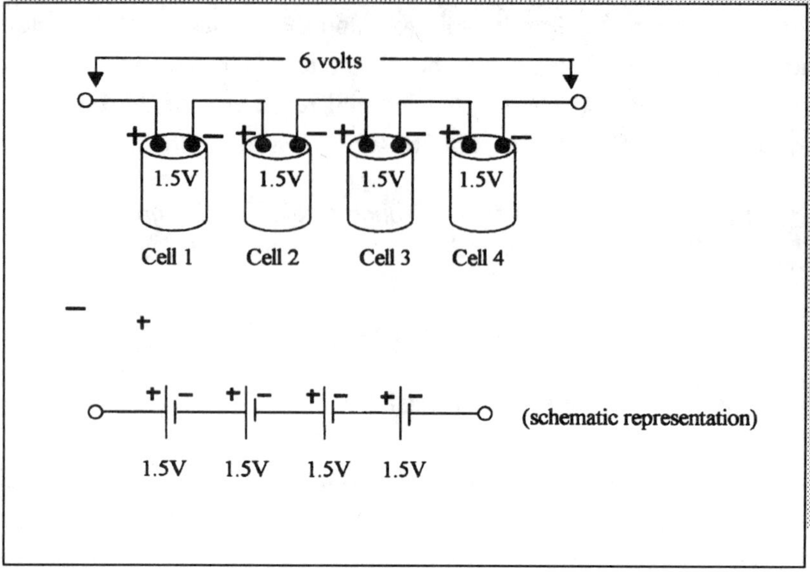

Figure 2.4
Cells in series.

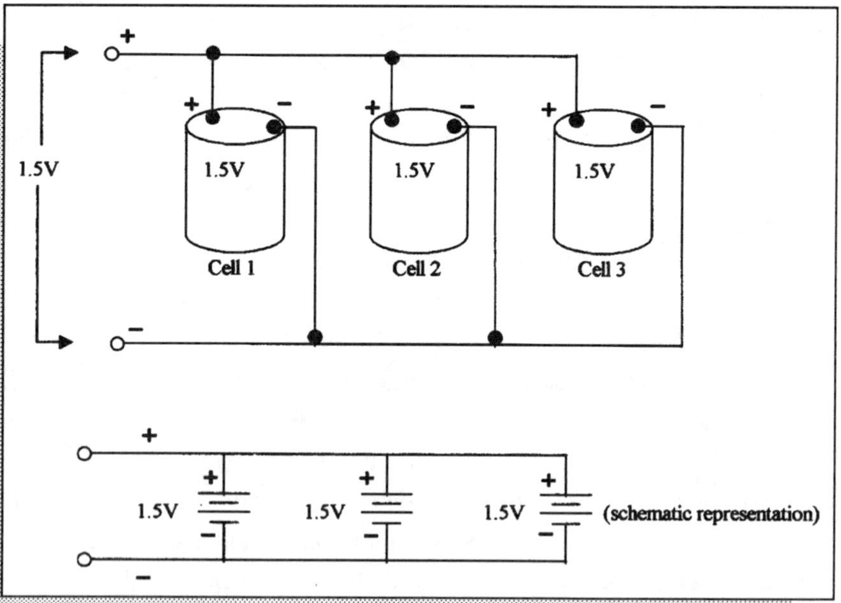

Figure 2.5
Cells in parallel.

The total voltage output of a battery of three parallel cells is the same as that for a single cell (Figure 2.5), but the available current is three times that of one cell; that is, the current capacity has been increased.

Identical cells in parallel all supply equal parts of the current to the load. For example, of three different parallel cells producing a load current of 210 ma, each cell contributes 70 ma.

Figure 2.6 depicts a schematic of a **series-parallel** battery network supplying power to a load requiring both a voltage and current greater than one cell can provide. To provide the required increased voltage, groups of three 1.5-V cells are connected in series. To provide the required increased amperage, four series groups are connected in parallel.

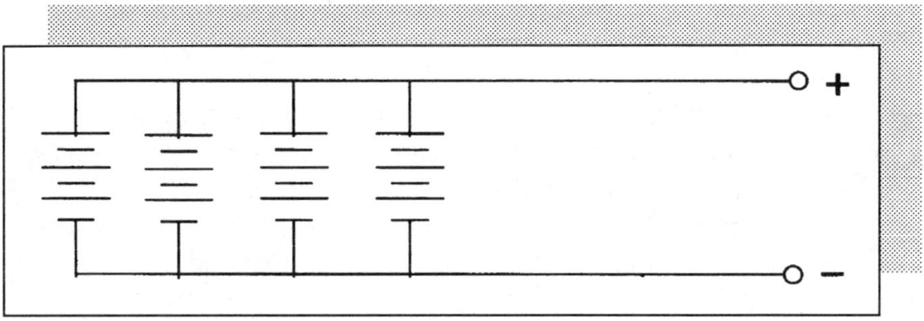

Figure 2.6
Series-parallel connected cells.

2.4 TYPES OF BATTERIES

In the past 25 years, several different types of batteries have been developed. In this text, we briefly discuss five types: the dry cell, lead-acid battery, alkaline cell, nickel-cadmium, and mercury cell.

2.4.1 DRY CELL

The dry cell, or carbon-zinc cell, is so called because its electrolyte is not in a liquid state (however, the electrolyte is a moist paste). The dry cell battery is one of the oldest and most widely used commercial types of dry cell. The carbon, in the form of a rod that is placed in the center of the cell, is the positive terminal. The case of the cell is made of zinc, which is the negative terminal (see Figure 2.2). Between the carbon electrode and the zinc case is the electrolyte of a moist chemical paste-like mixture. The cell is sealed to prevent the liquid in the paste from evaporating. The voltage of a cell of this type is about 1.5 V.

2.4.2 LEAD-ACID BATTERY

The *lead-acid battery* is a secondary cell—commonly termed a storage battery—that stores chemical energy until it is released as electrical energy.

IMPORTANT

Note: *The lead-acid battery differs from the primary cell type battery mainly in that it may be recharged, whereas most primary cells are not normally recharged. Also, the term "storage battery" is somewhat deceiving because this battery does not store electrical energy, but is a source of chemical energy that produces electrical energy.*

As the name implies, the lead-acid battery consists of a number of lead-acid cells immersed in a dilute solution of sulfuric acid. Each cell has two groups of lead plates; one set is the positive terminal, and the other is the negative terminal. Active materials within the battery (lead plates and sulfuric acid electrolyte) react chemically to produce a flow of direct current whenever current-consuming devices are connected to the battery terminal posts. This current is produced by chemical reaction between the active material of the plates (electrodes) and the electrolyte (sulfuric acid).

This type of cell produces slightly more than 2 V. Most automobile batteries contain six cells connected in series so that the output voltage from the battery is slightly more than 12 V.

Besides being rechargeable, the main advantage of the lead-acid storage battery over the dry cell battery is that the storage battery can supply current for a much longer time than the average dry cell.

IMPORTANT ***Safety Note:*** *Whenever a lead-acid storage battery is charging, the chemical action produces dangerous hydrogen gas; thus, the charging operation should only take place in a well-ventilated area.*

2.4.3 ALKALINE CELL

The *alkaline cell* is a secondary cell that gets its name from its alkaline electrolyte—potassium hydroxide. Another type of battery, sometimes called the "alkaline battery," has a negative electrode of zinc and a positive electrode of manganese dioxide. It generates 1.5 V.

2.4.4 NICKEL-CADMIUM CELL

The *nickel-cadmium cell,* or Ni-Cad cell, is the only dry battery that is a true storage battery with a reversible chemical reaction, allowing recharging many times. In the secondary nickel-cadmium dry cell, the electrolyte is potassium hydroxide, the negative electrode is nickel hydroxide, and the positive electrode is cadmium oxide. The operating voltage is 1.25 V.

Because of its rugged characteristics (stands up well to shock, vibration, and temperature changes) and availability in a variety of shapes and sizes, it is ideally suited for use in powering portable communication equipment.

2.4.5 MERCURY CELL

The *mercury cell* was developed as a result of space exploration activities, the development of small transceivers and miniaturized equipment where a power source of miniaturized size was needed. In addition to reduced size, the mercury cell has a good shelf life and is very rugged; it also produces a constant output voltage under different load conditions.

There are two different types of mercury cells. One is a flat cell that is shaped like a button, while the other is a cylindrical cell that looks like a standard flashlight cell. The advantage of the button-type cell is that several of them can be stacked inside one container to form a battery. A cell produces 1.35 V.

2.5 BATTERY CHARACTERISTICS

Batteries are generally classified by their various characteristics. Parameters such as internal resistance, specific gravity, capacity, and shelf life are used to classify batteries by type.

Regarding **internal resistance,** it is important to keep in mind that a battery is a d-c voltage generator. As such, the battery has internal resistance. In a chemical cell, the resistance of the electrolyte between the electrodes is responsible for most of the cell's internal resistance. Because any current in the battery must flow through the internal resistance, this resistance is in series with the generated voltage. With no current, the voltage drop across the resistance is zero so that the full generated voltage develops across the output terminals. This is the open-circuit voltage, or no-load voltage. If a load resistance is connected across the battery, the load resistance is in series with internal resistance. When current flows in this circuit, the internal voltage drop decreases the terminal voltage of the battery.

The ratio of the weight of a certain volume of liquid to the weight of the same volume of water is called the specific gravity of the liquid. Pure sulfuric acid has a specific gravity of 1.835 because it weighs 1.835 times as much as water per unit volume. The specific gravity of a mixture of sulfuric acid and water varies with the strength of the solution from 1.000 to 1.830.

The specific gravity of the electrolyte solution in a lead-acid cell ranges from 1.210 to 1.300 for new, fully charged batteries. The higher the **specific gravity,** the less internal resistance of the cell and the higher the possible load current. As the cell discharges, the water formed dilutes the acid, and the specific gravity gradually decreases to about 1.150, at which time the cell is considered to be fully discharged.

The specific gravity of the electrolyte is measured with a *hydrometer,* which has a compressible rubber bulb at the top, a glass barrel, and a rubber hose at the bottom of the barrel. In taking readings with a hydrometer, the decimal point is usually omitted. For example, a specific gravity of 1.260 is read simply as "twelve-sixty." A hydrometer reading of 1210 to 1300 indicates full charge; about 1250 is half-charge; and 1150 to 1200 is complete discharge.

The *capacity* of a battery is measured in ampere-hours (Ah).

IMPORTANT **Note:** *The ampere-hour capacity is equal to the product of the current in amperes and the time in hours during which the battery is supplying this current. The ampere-hour capacity varies inversely with the discharge current. The size of a cell is determined generally by its ampere-hour capacity.*

The capacity of a storage battery determines how long it will operate at a given discharge rate and depends upon many factors. The most important of these are as follows:

- the area of the plates in contact with the electrolyte
- the quantity and specific gravity of the electrolyte
- the type of separators
- the general condition of the battery (degree of sulfating, plates buckled, separators warped, sediment in bottom of cells, etc.)
- the final limiting voltage

The **shelf life** of a cell is that period of time during which the cell can be stored without losing more than approximately 10 percent of its original capacity. The loss of capacity of a stored cell is due primarily to the drying out of its electrolyte in a wet cell and to chemical actions that change the materials within the cell. The shelf life of a cell can be extended by keeping it in a cool, dry place.

Self-Test

2.1 What method of producing a voltage is used in batteries?

2.2 A ________________ consists of ______________ or more cells connected in series or parallel.

2.3 Current flows away from what electrode of a cell?

2.4 A cell or a battery is recharged by passing current through it in a direction ______________ to the direction of its discharge current.

2.5 What are the two basic components that cause a cell to produce electricity?

2.6 Cells that can be effectively recharged are called __________ cells.

2.7 In order to obtain a greater current capacity, cells are connected in _______________.

2.8 What is the major difference between a wet cell and a dry cell?

2.9 To increase both current and voltage capacity, cells are connected in _____________________.

2.10 What is the form of the electrolyte of a dry cell?

D-C Circuits

TOPICS

The Simple Electric Circuit
Ohm's Law
Electric Power
Electric Energy
Series D-C Circuit Characteristics
Parallel D-C Circuits
Series-Parallel Circuits

3.1 INTRODUCTION

In Chapters 1 and 2, we introduced some of the basic concepts of electricity. Now we will use those fundamental ideas along with the concepts contained in Ohm's Law to begin to understand the direct current (d-c) electric circuit.

3.2 THE SIMPLE ELECTRIC CIRCUIT

An electric circuit includes an **energy source** [source of electromotive force (emf) or voltage; that is, a battery or generator], a conductor (wire), a load, and a means of control (see Figure 3.1). The energy source could be a battery, as in Figure 3.1, or some other means of producing a voltage. The **load** that dissipates the energy could be a lamp, a resistor, or some other device (or devices) that does useful work, such as an electric toaster, a power drill, radio, or soldering iron. **Conductors** are wires that offer low resistance to current (see Chapter 4); they connect all the loads in the circuit to the voltage source. No electrical device dissipates energy unless current flows through it. Because conductors, or wires, are not perfect conductors, they heat up (dissipate energy), so they are actually part of the load. For simplicity, however, we usually think of the connecting

Key Terms Used in This Chapter

SCHEMATIC DIAGRAM	A simplified drawing that represents the electrical situation in a circuit.
OHM'S LAW	Defines the relationship between current, voltage, and resistance.
POWER	Pertains to the rate at which work is being done.
WORK	What is done whenever a force causes motion.
ENERGY	The ability to do work.
SERIES CIRCUIT	Arranged so that the electrons (current flow) have only one possible path.
KIRCHHOFF'S VOLTAGE LAW	States that the voltage applied to a closed circuit Equals the sum of the voltage drops in that circuit.
GROUND	Used to denote a common electrical point of zero potential.
PARALLEL CIRCUIT	A circuit having two or more components connected across the same voltage source.
KIRCHHOFF'S CURRENT LAW	States the algebraic sum of the currents entering and leaving any junction of conductors is equal to zero.
RECIPROCAL	An inverted fraction.
SERIES-PARALLEL CIRCUIT	A circuit containing both series and parallel elements.

wiring as having no resistance, because it would be tedious to assign a very low resistance value to the wires every time we wanted to solve a problem. **Control devices** might be switches, variable resistors, circuit breakers, fuses, or relays.

A complete pathway for current flow, or **closed circuit** (Figure 3.1), is an unbroken path for current from the emf, through a load, and back to the source. A circuit is called **open** (see Figure 3.2) if a break in the circuit (e.g., open switch) does not provide a complete path for current.

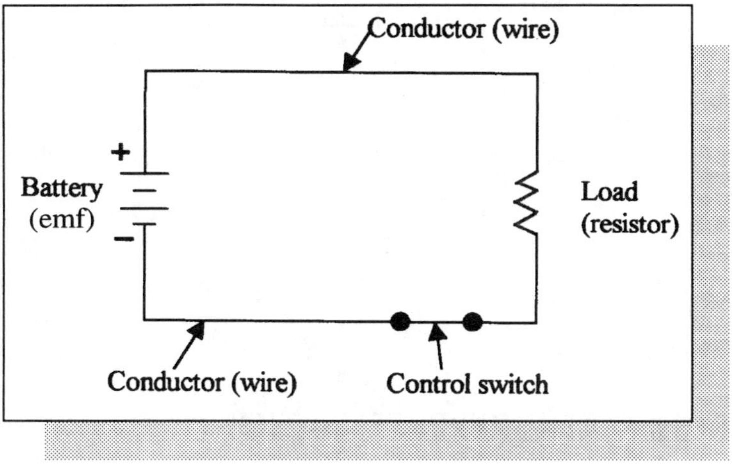

Figure 3.1
Simple closed circuit.

IMPORTANT ***Important Point:*** *Current flows from the negative (–) terminal of the battery, shown in Figures 3.1 and 3.2, through the load to the positive (+) battery terminal, and continues by going through the battery from the positive (+) terminal to the negative (–) terminal. As long as this pathway is unbroken, it is a closed circuit, and current will flow. However, if the path is broken at any point, it is an open circuit, and no current flows (see Figure 3.2).*

To protect a circuit, a fuse is placed directly into the circuit (see Figure 3.3). A fuse will open the circuit whenever a dangerous large current starts to flow (i.e., a short-circuit condition occurs, caused by an accidental connection between two points in a circuit that offer very little resistance). A fuse will permit currents smaller than the fuse value to flow but will melt and therefore break or open the circuit if a larger current flows.

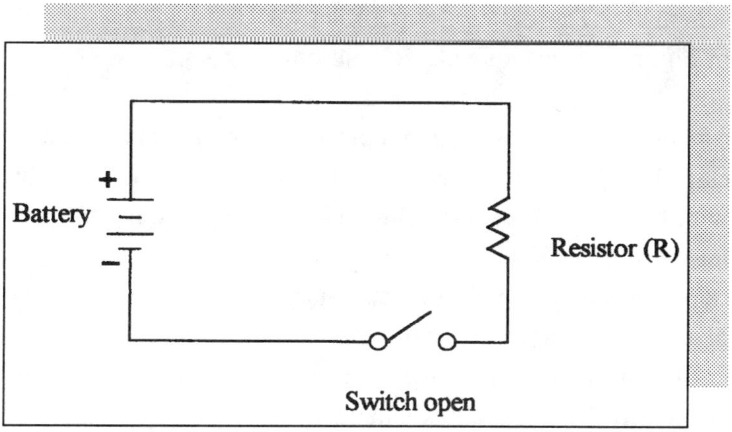

Figure 3.2
Open circuit.

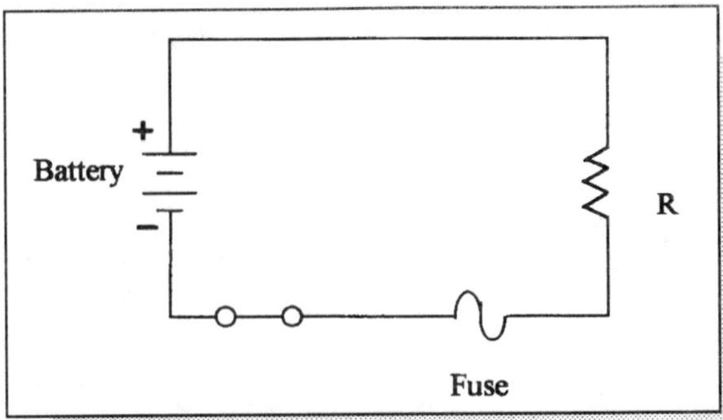

Figure 3.3
A simple fused circuit.

3.2.1 SCHEMATIC REPRESENTATION

The simple circuits shown in Figures 3.1, 3.2, and 3.3 are displayed in **schematic** form. A *schematic diagram* (usually shortened to "schematic") is a simplified drawing that represents the **electrical,** not the physical, situation in a circuit. The symbols used in schematic diagrams are the electrician's "shorthand"; they make the diagrams easier to draw and easier to understand. Consider the symbol in Figure 3.4 used to represent a battery power supply. The symbol is rather simple and straightforward—but is also very important. For example, by convention, the shorter line in the symbol for a battery represents the **negative** terminal. It is important to remember this, because it is sometimes necessary to note the direction of current flow, which is from negative to positive, when you examine the schematic. The battery symbol shown in Figure 3.4 has a single cell, so only one short and one long line are used. The number of lines used to represent a battery vary (and they are not necessarily equivalent to the number of cells), but they are always in pairs, with long and short lines alternating. In the circuit shown in Figure 3.3, the current would flow in a *counterclockwise* direction; that is, in the opposite direction that a clock's hands move. If the long and short lines of the battery symbol (symbol shown in Figure 3.4) were reversed, the current in the circuit shown in Figure 3.3 would flow *clockwise;* that is, in the direction of a clock's hands.

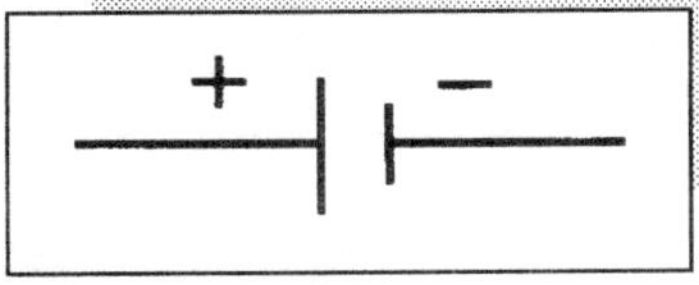

Figure 3.4
Schematic symbol for a battery.

IMPORTANT **Important Note:** *In studies of electricity and electronics, many circuits are analyzed that consist mainly of specially designed resistive components. As previously stated, these components are called* **resistors.** *Throughout the remaining analysis of the basic circuit, the resistive component will be a physical resistor. However, the resistive component could be any one of several electrical devices.*

Keep in mind that in the simple circuits shown in the figures to this point, we have only illustrated and discussed a few of the many symbols used in schematics to represent circuit components. (Other symbols will be introduced as we need them.)

It is also important to keep in mind that a closed loop of wire (conductor) is not necessarily a circuit. A source of voltage must be included to make it an electric circuit. In any electric circuit where electrons move around a closed loop, current, voltage, and resistance are present. The physical pathway for current flow is actually the circuit. By knowing any two of the three quantities, such as voltage and current, the third (resistance) may be determined. This is done mathematically using Ohm's Law, which is the foundation on which electrical theory is based.

3.3 OHM'S LAW

Simply put, *Ohm's Law* defines the relationship between current, voltage, and resistance in electric circuits. Ohm's Law can be expressed mathematically in three ways.

a. The *current* in a circuit is equal to the voltage applied to the circuit divided by the resistance of the circuit. Stated another way, the current in a circuit is *directly* proportional to the applied voltage and *inversely* proportional to the circuit resistance. Ohm's Law may be expressed as an equation:

$$I = \frac{E}{R} \tag{3.1}$$

where

I = current in amperes (amps)
E = voltage in volts
R = resistance in ohms

b. The *resistance* of a circuit is equal to the voltage applied to the circuit divided by the current in the circuit:

$$R = \frac{E}{I} \tag{3.2}$$

c. The applied *voltage* (***E***) to a circuit is equal to the product of the current and the resistance of the circuit:

$$E = I \times R = IR \tag{3.3}$$

If any two of the quantities in Equations (3.1–3.3) are known, the third may be easily found. Let's look at an example.

EXAMPLE 3.1

Problem:

Figure 3.5 shows a circuit containing a resistance of 6 ohms and a source of voltage of 3 volts. How much current flows in the circuit?

Given:

$E = 3$ volts
$R = 6$ ohms
$I = ?$

Solution:

$$I = \frac{E}{R} \tag{3.1}$$

$$I = \frac{3}{6}$$

$I = 0.5$ amperes

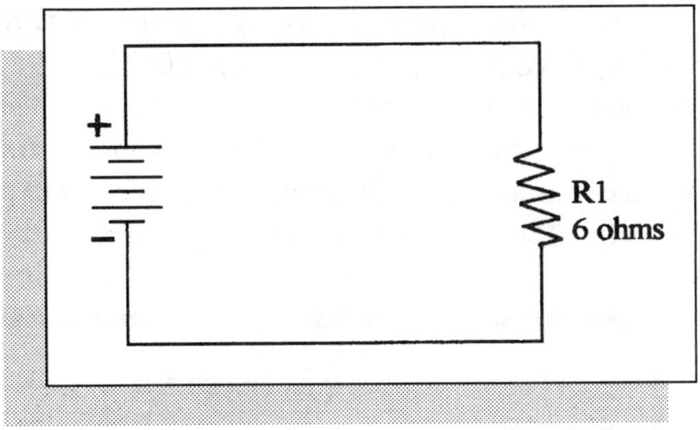

Figure 3.5
Determining current in a simple circuit.

To observe the effect of source voltage on circuit current, we use the circuit shown in Figure 3.5 but double the voltage to 6 volts.

Notice that as the source of voltage doubles, the circuit current also doubles.

EXAMPLE 3.2

Problem:

Given:

$E = 6$ volts
$R = 6$ ohms
$I = ?$

Solution:

$$I = \frac{E}{R} \qquad (3.1)$$

$$I = \frac{6}{6}$$

$I = 1$ ampere

IMPORTANT ***The Key Point:*** *Circuit current is directly proportional to applied voltage and will change by the same factor that the voltage changes.*

To verify that current is inversely proportional to resistance, assume the resistor in Figure 3.5 to have a value of 12 ohms.

Comparing this current of 0.25 ampere for the 12-ohm resistor to the 0.5-ampere of current obtained with the 6-ohm resistor shows that doubling the resistance will reduce the current to one-half the original value. **The point:** Circuit current is inversely proportional to the circuit resistance.

EXAMPLE 3.3

Problem:

Given:

$E = 3$ volts

$R = 12$ ohms

$I = ?$

Solution:

$$I = \frac{E}{R} \quad (3.1)$$

$$I = \frac{3}{12}$$

$I = 0.25$ amperes

Recall that if you know any two quantities (*E, I,* and *R*), you can calculate the third. In many circuit applications, current is known, and either the voltage or the resistance will be the unknown quantity. To solve a problem in which current and resistance are known, the basic formula for Ohm's Law must be transposed to solve for *E*, for *I*, or for *R*.

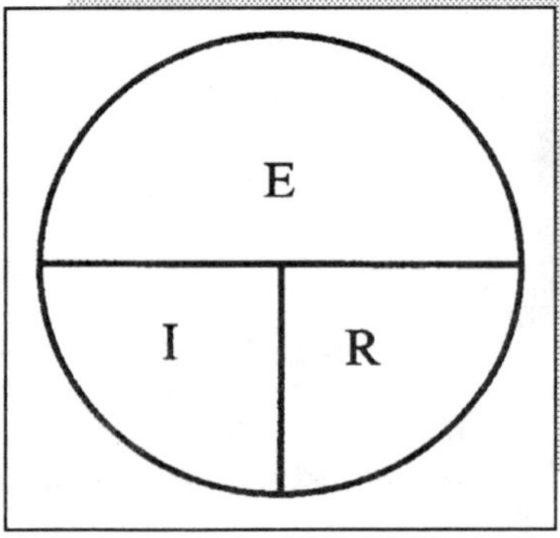

Figure 3.6
Ohm's Law circle.

However, the Ohm's Law equations can be memorized and practiced effectively by using an Ohm's Law circle (see Figure 3.6).

To find the equation for *E*, *I*, or *R* when two quantities are known, cover the unknown third quantity with your finger, ruler, piece of paper, etc., as shown in Figure 3.7.

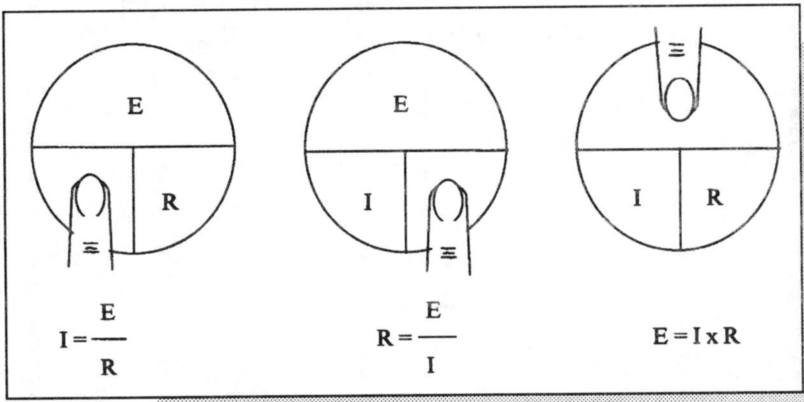

Figure 3.7
Putting the Ohm's Law circle to work.

EXAMPLE 3.4

Problem: Find *I* when *E* = 120 V and *R* = 40 ohms (Ω).

Solution:
Place finger on *I* as shown in the figure below.
Use Equation (3.1) to find the unknown *I*.

$$I = \frac{E}{R} \tag{3.1}$$

$$= \frac{120}{40} = 3\ \text{A}$$

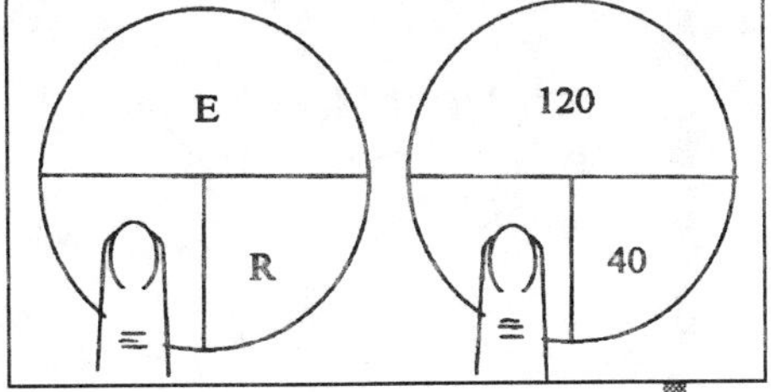

EXAMPLE 3.5

Problem: Find R when $E = 220$ V and $I = 10$ A

Solution: Place finger on R as shown in the figure. Use Equation (3.2) to find the unknown R.

$$R = \frac{E}{I}$$

$$= \frac{220}{10} = 22\ \Omega$$

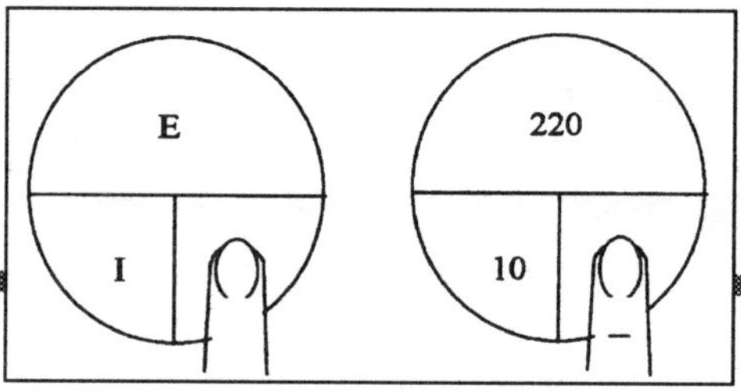

EXAMPLE 3.6

Problem: Find E when $I = 2.5$ A and $R = 25\ \Omega$.

Solution: Place finger on E as shown in the figure.

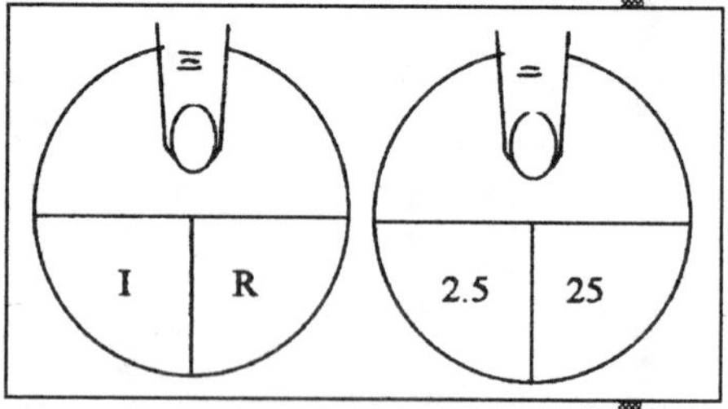

$$E = IR = 2.5(25) = 62.5\ V$$

IMPORTANT

Note: *In the previous examples, we have demonstrated how the Ohm's Law circle can help us solve simple voltage, current, and amperage problems. The beginning student is cautioned, however, not to rely wholly on the use of this circle when transposing simple formulas but rather to use it to supplement his or her knowledge of the algebraic method. Algebra is a basic tool in the solution of electrical problems, and the importance of knowing how to use it should not be under-emphasized or bypassed after the operator has learned a shortcut method such as the one indicated in this circle.*

EXAMPLE 3.7

Problem: An electric light bulb draws 0.5 A when operating on a 120-V d-c circuit. What is the resistance of the bulb?

Solution: The first step in solving a circuit problem is to sketch a schematic diagram of the circuit itself, labeling each of the parts and showing the known values (see Figure 3.8).

Because I and E are known, we use Equation (3.2) to solve for R.

$$R = \frac{E}{I} = \frac{120}{0.5} = 240\ \Omega \qquad (3.2)$$

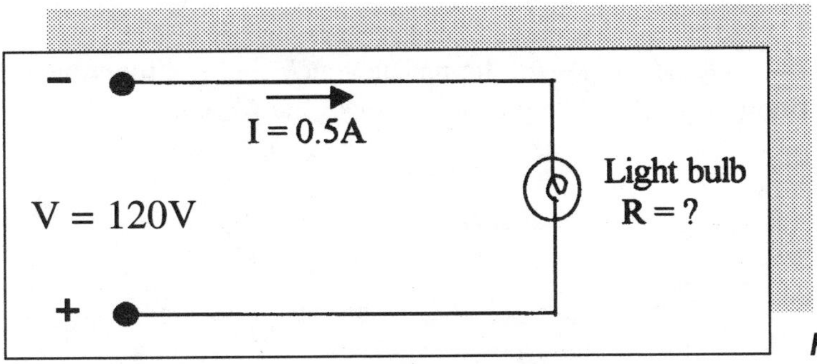

Figure 3.8.

3.4 ELECTRIC POWER

Power, whether electrical or mechanical, pertains to the rate at which work is being done, so the power consumption in your plant is related to current flow. A large electric motor or air dryer consumes more power (and draws more current) in a given length of time than, for example, an indicating light on a motor controller. **Work** is done whenever a force causes motion. If a mechanical force is used to lift or move a weight, work is done. However, force exerted *without* causing motion, such as the force of a compressed spring acting between two fixed objects, does not constitute work.

IMPORTANT

Key Point: *Power is the rate at which work is done.*

3.4.1 ELECTRICAL POWER CALCULATIONS

The electric power P used in any part of a circuit is equal to the current I in that part multiplied by the V across that part of the circuit. In equation form,

$$P = EI \tag{3.4}$$

where

P = power, watts (w)
E = voltage, V
I = current, A

If we know the current (I) and the resistance (R) but not the voltage (V), we can find the power (P) by using Ohm's Law for voltage, so that substituting

$$E = IR \tag{3.3}$$

into Equation (3.4) we have

$$P = IR \times I = I^2R \tag{3.5}$$

In the same manner, if we know the voltage V and the resistance R but not the current I, we can find the P by using Ohm's Law for current, so that substituting

$$I = \frac{E}{R} \tag{3.1}$$

into Equation (3.4) we have

$$P = E\frac{E}{R} = \frac{E^2}{R} \tag{3.6}$$

IMPORTANT

Key Point: *If we know any two quantities, we can calculate the third.*

EXAMPLE 3.8

Problem: The current through a 200-Ω resistor to be used in a circuit is 0.25 A. Find the power rating of the resistor.

Solution: Because I and R are known, use Equation (3.5) to find P.

$$P = I^2R = (0.25)^2(200) = 0.0625(200) = 12.5 \text{ w}$$

IMPORTANT

Important Point: *The power rating of any resistor used in a circuit should be twice the wattage calculated by the power equation to prevent the resistor from burning out. Thus, the resistor used in Example 3.8 should have a power rating of 25 watts.*

EXAMPLE 3.9

Problem: How many kilowatts of power are delivered to a circuit by a 220-V generator that supplies 30 A to the circuit?

Solution: Because *V* and *I* are given, use Equation (3.4) to find *P*.

$$P = EI = 220(30) = 6600 \text{ w} = 6.6 \text{ kw}$$

In this section, *P* was expressed in terms of alternate pairs of the other three basic quantities *E, I,* and *R.* In practice, you should be able to express any one of the three basic quantities, as well as *P*, in terms of any two of the others. Figure 3.9 is a summary of 12 basic formulas you should know. The four quantities *E, I, R,* and *P* are at the center of the figure.

Adjacent to each quantity are three segments. Note that in each segment, the basic quantity is expressed in terms of two other basic quantities, and no two segments are alike.

EXAMPLE 3.10

Problem: If the voltage across a 30,000-Ω resistor is 450 V, what is the power dissipated in the resistor?

Solution: Because *R* and *E* are known, use Equation (3.6) to find *P.*

$$P = \frac{E^2}{R} = \frac{450^2}{30,000} = \frac{202,500}{30,000} = 6.75 \text{ w}$$

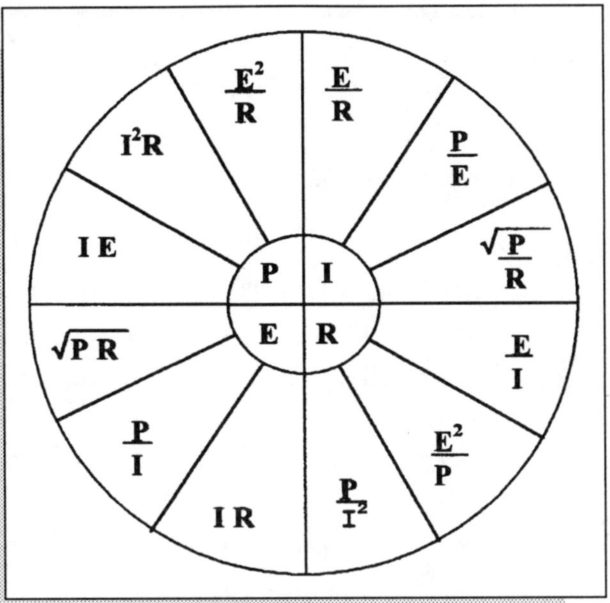

Figure 3.9 *Ohm's Law circle—summary of basic formulas.*

3.5 ELECTRIC ENERGY

Energy (the mechanical definition) is defined as the ability to do work (energy and time are essentially the same and are expressed in identical units). Energy is expended when work is done, because it takes energy to maintain a force when that force acts through a distance. The total energy expended to do a certain amount of work is equal to the working force multiplied by the distance through which the force moved to do the work.

In electricity, total energy expended is equal to the *rate* at which work is done, multiplied by the length of time the rate is measured. Essentially, energy (*W*) is equal to power (*P*) times time (*t*).

The kilowatt-hour (kwh) is a unit commonly used for large amounts of electric energy or work. The amount of kilowatt-hours is calculated as the product of the power in kilowatts (kw) and the time in hours (h) during which the power is used.

$$\mathbf{kwh = kw \times h} \qquad (3.7)$$

EXAMPLE 3.11

Problem: How much energy is delivered in 4 hours by a generator supplying 12 kw?

Solution:

$$\text{kwh} = \text{kw} \times \text{h} \qquad (3.7)$$

$$= 12(4) = 48$$

$$\text{Energy delivered} = 48 \text{ kwh}$$

3.6 SERIES D-C CIRCUIT CHARACTERISTICS

As previously mentioned, an electric circuit is made up of a voltage source, the necessary connecting conductors, and the effective load.

If the circuit is arranged so that the electrons have only *one* possible path, the circuit is called a *series circuit.* Therefore, a series circuit is defined as a circuit that contains only one path for current flow. Figure 3.10 shows a series circuit having several loads (resistors).

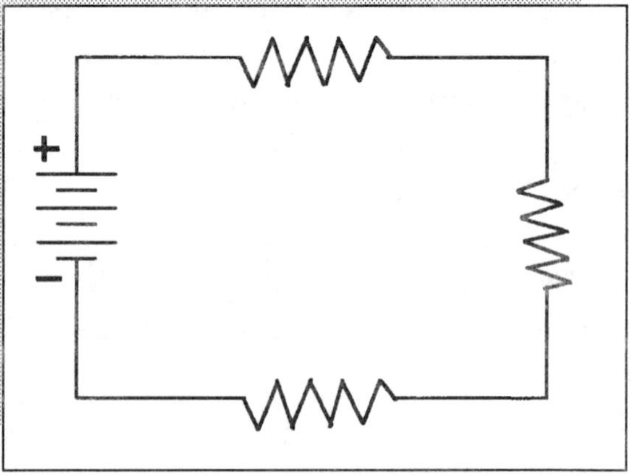

Figure 3.10
Series circuit.

IMPORTANT

Key Point: *A* series circuit *is a circuit in which there is only one path for current to flow along.*

3.6.1 RESISTANCE

Referring to Figure 3.10, the current in a series circuit, in completing its electrical path, must flow through each resistor inserted into the circuit. Thus, each additional resistor offers added resistance. In a series circuit, *the total circuit resistance (R_T) is equal to the sum of the individual resistances.* As an equation:

$$R_T = R_1 + R_2 + R_3 \ldots R_n \qquad (3.8)$$

where

R_T = total resistance, Ω
R_1, R_2, R_3 = resistance in series, Ω
R_n = any number of additional resistors in equation

IMPORTANT

Key Point: *When resistances are connected in series, the total resistance in the circuit is equal to the sum of the resistances of all the parts of the circuit.*

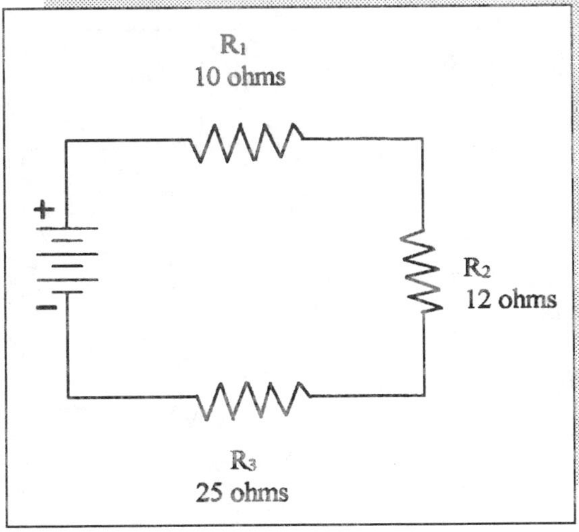

Figure 3.11
Solving for total resistance in a series circuit.

EXAMPLE 3.12

Problem: Three resistors of 10 ohms, 12 ohms, and 25 ohms are connected in series across a battery whose emf is 110 volts (Figure 3.11). What is the total resistance?

Solution:

Given: $R_1 = 10$ ohms
$R_2 = 12$ ohms
$R_3 = 25$ ohms
$R_T = ?$

$$R_T = R_1 + R_2 + R_3$$
$$R_T = 10 + 12 + 25$$
$$R_T = 47 \text{ ohms}$$

Equation (3.8) can be transposed to solve for the value of an unknown resistance. For example, transposition can be used in some circuit applications where the total resistance is known but the value of a circuit resistor has to be determined.

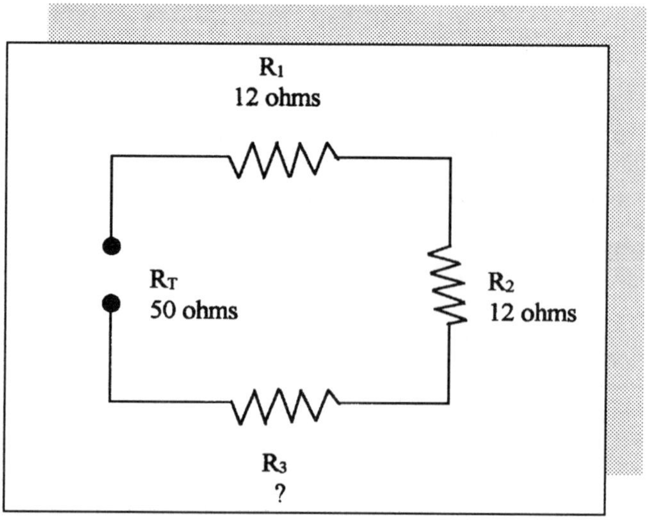

Figure 3.12 *Calculating the value of one resistance in a series circuit.*

EXAMPLE 3.13

Problem: The total resistance of a circuit containing three resistors is 50 ohms (see Figure 3.12). Two of the circuit resistors are 12 ohms each. Calculate the value of the third resistor.

Solution:

Given:

$$R_T = 50 \text{ ohms}$$
$$R_1 = 12 \text{ ohms}$$
$$R_2 = 12 \text{ ohms}$$
$$R_3 = ?$$
$$R_T = R_1 + R_2 + R_3$$

Subtracting $(R_1 + R_2)$ from both sides of the equation

$$R_3 = R_T - R_1 - R_2$$
$$R_3 = 50 - 12 - 12$$
$$R_3 = 50 - 24$$
$$R_3 = 26 \text{ ohms}$$

3.6.2 CURRENT

Because there is but one path for current in a series circuit, the same *current* (I) must flow through each part of the circuit. Thus, to determine the current throughout a series circuit, only the current through one of the parts need be known.

The fact that the same current flows through each part of a series circuit can be verified by inserting ammeters into the circuit at various points as shown in Figure 3.13. As indicated in Figure 3.13, each meter indicates the same value of current.

IMPORTANT ***Key Point:*** *In a series circuit, the same current flows in every part of the circuit.* ***Do not*** *add the currents in each part of the circuit to obtain* I.

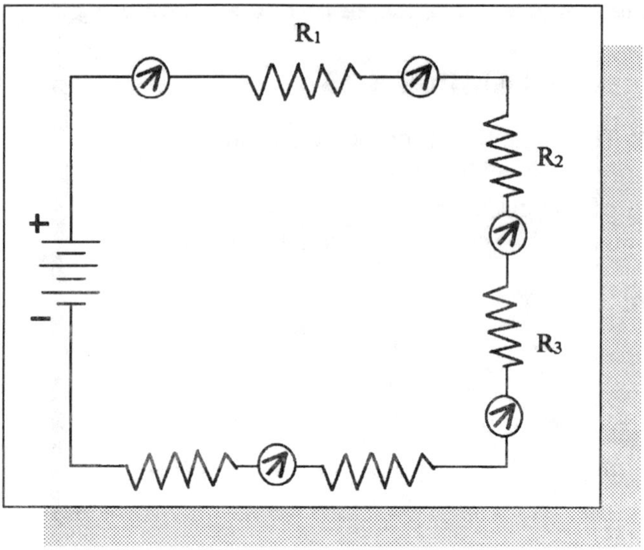

Figure 3.13
Current in a series circuit.

3.6.3 VOLTAGE

The *voltage* drop across the resistor in the basic circuit is the total voltage across the circuit and is equal to the applied voltage. The total voltage across a series circuit is also equal to the applied voltage, but consists of the sum of two or more individual voltage drops. This statement can be proven by an examination of the circuit shown in Figure 3.14.

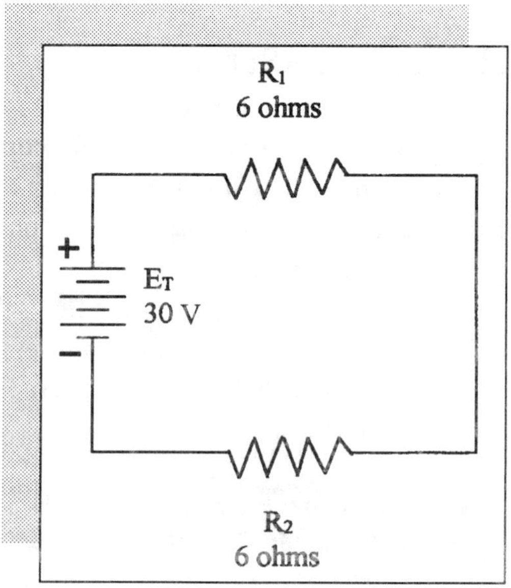

Figure 3.14
Calculating total resistance in a series circuit.

In this circuit, a source potential (E_T) of 30 volts is impressed across a series circuit consisting of two 6-ohm resistors. The total resistance (R_T) of the circuit is equal to the sum of the two individual resistances, or 12 ohms. Using Ohm's Law, the circuit current may be calculated as follows:

$$I = \frac{E_T}{R_T} \tag{3.1}$$

$$I = \frac{30}{12}$$

$$I = 2.5 \text{ amperes}$$

Knowing the value of the resistors to be 6 ohms each, and the current through the resistors to be 2.5 amperes, the voltage drops across the resistors can be calculated. The voltage (E_1) across R_1 is therefore

$$E_1 = IR_1$$
$$E_1 = 2.5 \text{ amps} \times 6 \text{ ohms}$$
$$E_1 = 15 \text{ volts}$$

Because R_2 is the same ohmic value as R_1 and carries the same current, the voltage drop across R_2 is also equal to 15 volts. Adding these two 15-volt drops together gives a total drop of 30 volts, exactly equal to the applied voltage. For a series circuit then,

$$E_T = E_1 + E_2 + E_3 \ldots E_n \tag{3.9}$$

where

E_T = total voltage, V
E_1 = voltage across resistance R_1, V
E_2 = voltage across resistance R_2, V
E_3 = voltage across resistance R_3, V

EXAMPLE 3.14

Problem: A series circuit consists of three resistors having values of 10 ohms, 20 ohms, and 40 ohms, respectively. Find the applied voltage if the current through the 20-ohm resistor is 2.5 amperes.

Solution: To solve this problem, a circuit diagram is first drawn and labeled as shown in Figure 3.15.

Given:

$$R_1 = 10 \text{ ohms}$$
$$R_2 = 20 \text{ ohms}$$
$$R_3 = 40 \text{ ohms}$$
$$I = 2.5 \text{ amps}$$

Because the circuit involved is a series circuit, the same 2.5 amperes of current flows through each resistor. Using Ohm's Law, the voltage drops across each of the three resistors can be calculated and are

$$E_1 = 10 \times 2.5 = 25 \text{ volts}$$
$$E_2 = 20 \times 2.5 = 50 \text{ volts}$$
$$E_3 = 40 \times 2.5 = 100 \text{ volts}$$

Once the individual drops are known, they can be added to find the total or applied voltage using Equation (3.9):

$$E_T = E_1 + E_2 + E_3$$
$$E_T = 25\text{ V} + 50\text{ V} + 100\text{ V} \qquad (3.9)$$
$$E_T = 175 \text{ volts}$$

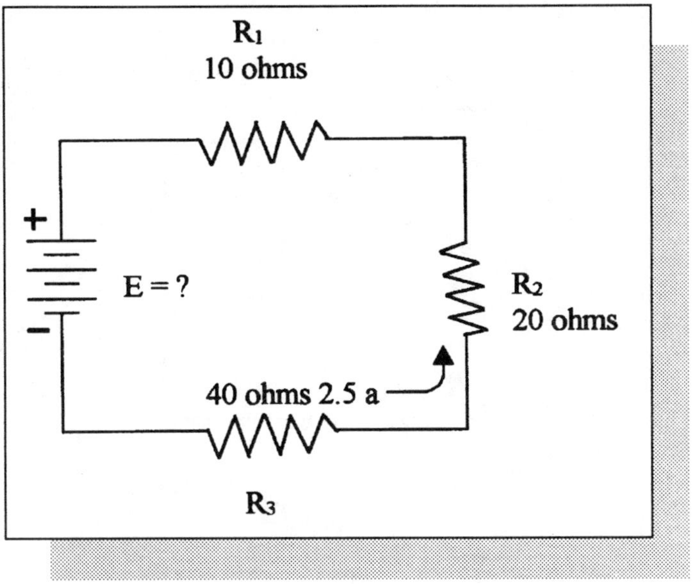

Figure 3.15
Solving for applied voltage in a series circuit.

IMPORTANT

Key Point 1: *The total voltage (E_T) across a series circuit is equal to the sum of the voltages across each resistance of the circuit.*

IMPORTANT

Key Point 2: *The voltage drops that occur in a series circuit are in direct proportions to the resistance across which they appear. This is the result of having the same current flow through each resistor. Thus, the larger the resistor, the larger will be the voltage drop across it.*

3.6.4 POWER

Each resistor in a series circuit consumes *power*. This power is dissipated in the form of heat. Because this power must come from the source, the total power must be equal in amount to the power consumed by the circuit resistances. In a series circuit, the total power is equal to the **sum** of the powers dissipated by the individual resistors. Total power (P_T) is thus equal to

$$P_T = P_1 + P_2 + P_3 \ldots P_n \qquad (3.10)$$

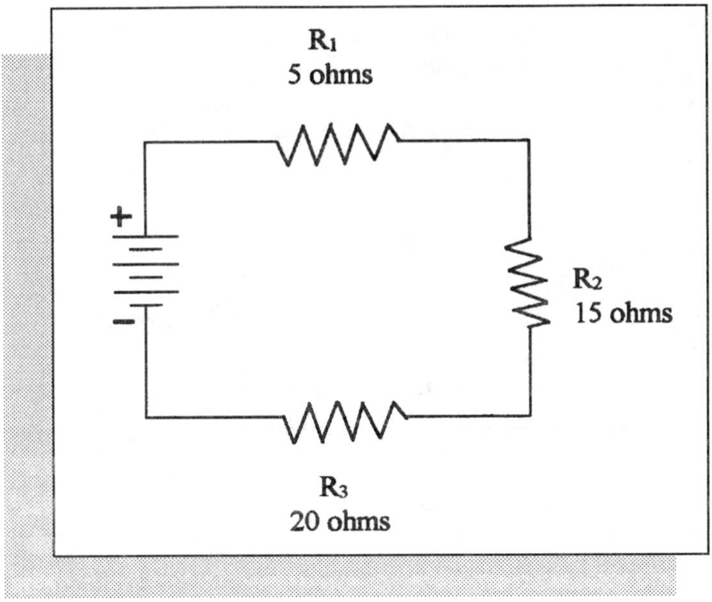

Figure 3.16 Solving for total power in a series circuit.

where

P_T = total power, w
P_1 = power used in first part, w
P_2 = power used in second part, w
P_3 = power used in third part, w
P_n = power used in *n*th part, w

EXAMPLE 3.15

Problem: A series circuit consists of three resistors having values of 5 ohms, 15 ohms, and 20 ohms. Find the total power dissipation when 120 volts is applied to the circuit (see Figure 3.16).

Solution:

Given:

$R_1 = 5$ ohms
$R_2 = 15$ ohms
$R_3 = 20$ ohms
$E = 120$ volts

The total resistance is found first.

$$R_T = R_1 + R_2 + R_3$$
$$R_T = 5 + 15 + 20 \qquad (3.8)$$
$$R_T = 40 \text{ ohms}$$

Using total resistance and the applied voltage, the circuit current is calculated.

$$I = \frac{E_T}{R_T}$$

$$I = \frac{120}{40}$$
$$I = 3 \text{ amps}$$

Using the power formula, the individual power dissipations can be calculated.

For resistor R_1:

$$P_1 = I^2R_1$$
$$P_1 = (3)^2 5$$
$$P_1 = 45 \text{ watts} \qquad (3.5)$$

For R_2:

$$P_2 = I^2R_2$$
$$P_2 = (3)^2 15$$
$$P_2 = 135 \text{ watts} \qquad (3.5)$$

For R_3:

$$P_3 = I^2R_3$$
$$P_3 = (3)^2 20$$
$$P_3 = 180 \text{ watts}$$

To obtain total power:

$$P_T = P_1 + P_2 + P_3$$
$$P_T = 45 + 135 + 180 \qquad (3.10)$$
$$P_T = 360 \text{ watts}$$

To check the answer, the total power delivered by the source can be calculated:

$$P = E \times I$$
$$P = 120 \text{ V} \times 3 \text{ a} \qquad (3.4)$$
$$P = 360 \text{ watts}$$

Thus, the total power is equal to the sum of the individual power dissipations.

$$P_T = IE_T \qquad (3.11)$$

IMPORTANT **Key Point:** *We found that Ohm's Law can be used for total values in a series circuit as well as for individual parts of the circuit. Similarly, the formula for power may be used for total values.*

3.6.5 SUMMARY OF THE RULES FOR SERIES D-C CIRCUITS

To this point, we have covered many of the important factors governing the operation of basic series circuits. In essence, what we have really done is to lay a strong foundation to build upon in preparation for more advanced circuit theory that follows. A summary of the important factors governing the operation of a series circuit are listed as follows:

- The same current flows through each part of a series circuit.
- The total resistance of a series circuit is equal to the sum of the individual resistances.

- The total voltage across a series circuit is equal to the sum of the individual voltage drops.
- The voltage drop across a resistor in a series circuit is proportional to the size of the resistor.
- The total power dissipated in a series circuit is equal to the sum of the individual dissipations.

3.6.6 GENERAL SERIES CIRCUIT ANALYSIS

Now that we have discussed the pieces involved in putting together the puzzle for solving series circuit analysis, we move on to the next step in the process: solving series circuit analysis in total.

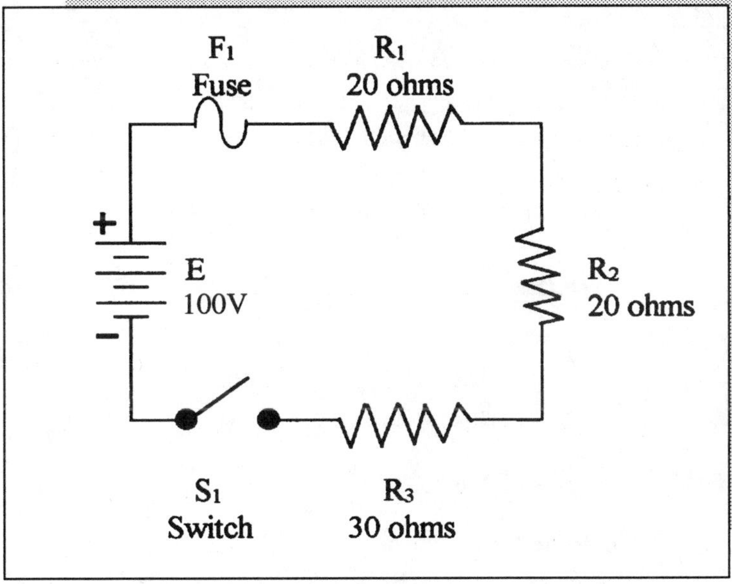

Figure 3.17
Solving for various values in a series circuit.

IMPORTANT **Note:** *In solving the circuit, the total resistance will be found first. Next, the circuit current will be calculated. Once the current is known, the voltage drops, and power dissipations can be calculated.*

EXAMPLE 3.16

Problem: Three resistors of 20 ohms, 20 ohms, and 30 ohms are connected across a battery supply rated at 100 volts terminal voltage. Completely solve the circuit shown in Figure 3.17.

Solution:

The total resistance is:

$$R_T = R_1 + R_2 + R_3$$
$$R_T = 20 \text{ ohms} + 20 \text{ ohms} + 30 \text{ ohms}$$
$$R_T = 70 \text{ ohms}$$

By Ohm's Law, the current is

$$I = \frac{E}{R_T}$$

$$I = \frac{100}{70}$$

$$I = 1.43 \text{ amps (rounded)}$$

The voltage (E_1) across R_1 is

$$E_1 = IR_1$$
$$E_1 = 1.43 \text{ amps} \times 20 \text{ ohms}$$
$$E_1 = 28.6 \text{ volts}$$

The voltage (E_2) across R_2 is

$$E_2 = IR_2$$
$$E_2 = 1.43 \text{ amps} \times 20 \text{ ohms}$$
$$E_2 = 28.6 \text{ volts}$$

The voltage (E_3) across R_3 is

$$E_3 = IR_2$$

$E_3 = 1.43 \text{ amps} \times 30 \text{ ohms}$
$E_3 = 42.9 \text{ volts}$

The power dissipated by R_1 is

$P_1 = I \times E_1$
$P_1 = 1.43 \text{ amps} \times 28.6 \text{ volts}$
$P_1 = 40.9 \text{ watts}$

The power dissipated by R_2 is

$P_2 = I \times E_2$
$P_2 = 1.43 \text{ amps} \times 28.6 \text{ volts}$
$P_2 = 40.9 \text{ watts}$

The power dissipated by R_3 is

$P_3 = I \times E_3$
$P_3 = 1.43 \text{ amps} \times 42.9 \text{ volts}$
$P_3 = 61.3 \text{ watts (rounded)}$

The total power dissipated is

$P_T = E_T \times I$
$P_T = 100 \text{ volts} \times 1.43 \text{ amps}$
$P_T = 143 \text{ watts}$

IMPORTANT

Important Note: *Keep in mind when applying Ohm's Law to a series circuit to consider whether the values used are component values or total values. When the information available enables the use of Ohm's Law to find total resistance, total voltage, and total current, total values must be inserted into the formula.*

To find total resistance: $R_T = \dfrac{E_T}{I_T}$

To find total voltage: $E_T = I_T \times R_T$

To find total current: $I_T = \dfrac{E_T}{R_T}$

3.6.7 KIRCHHOFF'S VOLTAGE LAW

Kirchhoff's voltage law states that *the voltage applied to a closed circuit equals the sum of the voltage drops in that circuit.* It should be obvious that this fact was used in the study of series circuits to this point. It was expressed as follows:

Voltage applied = sum of voltage drops

$$E_A = E_1 + E_2 + E_3 \tag{3.9}$$

where E_A is the applied voltage and E_1, E_2, and E_3 are voltage drops.

Another way of stating Kirchhoff's Law is that the algebraic sum of the instantaneous emf's and voltage drops around any closed circuit is zero.

Through the use of Kirchhoff's Law, circuit problems can be solved that would be difficult and often impossible with only knowledge of Ohm's Law. When Kirchhoff's Law is properly applied, an equation can be set up for a closed loop, and the unknown circuit values may be calculated.

3.6.8 POLARITY OF VOLTAGE DROPS

When there is a voltage drop across a resistance, one end must be more positive or more negative than the other end. The polarity of the voltage drop is determined by the direction of current flow. In the circuit shown in Figure 3.18, the current is seen to be flowing in a counterclockwise direction due to the arrangement of the battery source *E*. Notice that the end of resistor R_1 into which the current flows is marked **negative** (–). The end of R_1 at which the current leaves is marked **positive** (+). These polarity markings are used to show that the end of R_1 into which the current flows is at a higher negative potential than is the end of the resistor at which the current leaves. Point A is thus more negative than point B.

Point C, which is at the same potential as point B, is labeled negative. This is to indicate that point C, though positive with respect to point A, is more negative than point D. To say a point is positive (or negative), without stating what it is positive **in respect to,** has no meaning.

Kirchhoff's Voltage Law can be written as an equation as shown below:

$$E_a + E_b + E_c + \ldots E_n = 0 \qquad (3.12)$$

where E_a, E_b, etc., are the voltage drops and emfs around any closed circuit loop.

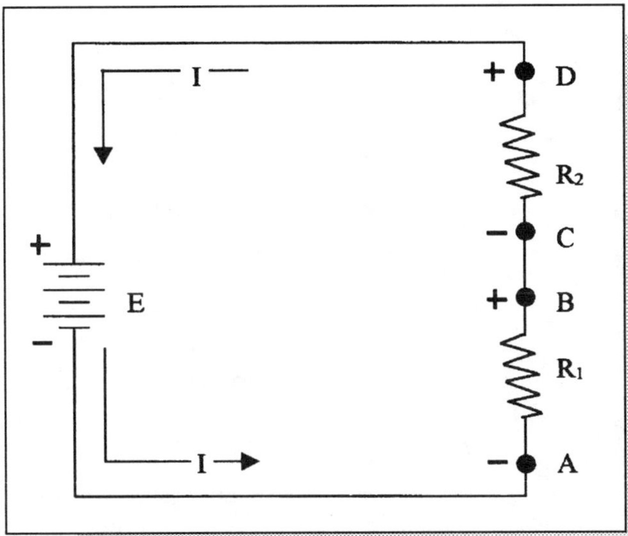

Figure 3.18
Polarity of voltage drops.

EXAMPLE 3.17

Problem: Three resistors are connected across a 60-volt source. What is the voltage across the third resistor if the voltage drops across the first two resistors are 10 volts and 20 volts?

Solution: First, draw a diagram like the one shown in Figure 3.19. Next, a direction of current is assumed as shown. Using this current, the polarity markings are placed at each end of each resistor and also on the terminals of the source. Starting at point A, trace around the circuit in the direction of current flow, recording the voltage and polarity of each component. Starting at point A, these voltages would be as follows:

Basic formula:

$$E_a + E_b + E_c \ldots E_n = 0 \qquad (3.12)$$

From the circuit:

$$(+E_?) + (+E_2) + (+E_3) - (E^A) = 0$$

Substituting values from circuit:

$$E_? + 10 + 20 - 60 = 0$$
$$E_? - 30 = 0$$
$$E_? = 30 \text{ volts}$$

Thus, the unknown voltage ($E_?$) is found to be 30 volts.

IMPORTANT

Note: *Using the same idea as above, a problem can be solved in which the current is the unknown quantity.*

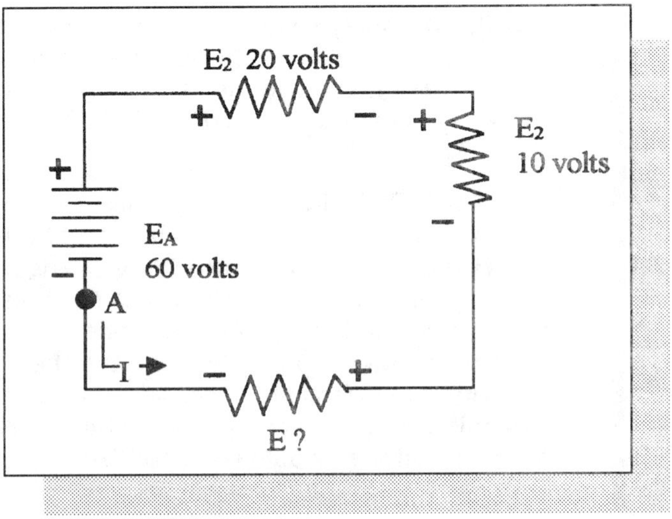

Figure 3.19
Determining unknown voltage in a series circuit.

3.6.9 SERIES AIDING AND OPPOSING SOURCES

Sources of voltage that cause current to flow in the same direction are considered to be *series aiding,* and their voltages add. Sources of voltage that would tend to force current in opposite directions are said to be *series opposing,* and the effective source voltage is the difference between the opposing voltages. When two opposing sources are inserted into a circuit, current flow would be in a direction determined by the larger source. Examples of series aiding and opposing sources are shown in Figure 3.20.

3.6.10 KIRCHHOFF'S LAW AND MULTIPLE SOURCE SOLUTIONS

Kirchhoff's Law can be used to solve multiple source circuit problems. In applying this method, the exact same procedure is used for the multiple source as was used for single source circuits. This is demonstrated by the following example.

EXAMPLE 3.18

Problem: Find the amount of current in the circuit shown in Figure 3.21.

Solution: Start at point A.

Basic equation:

$$E_a + E_b + E_c + \ldots E_n = 0 \qquad (3.12)$$

From the circuit:

$$E_{b2} + E_1 - E_{b1} + E_{b3} + E_2 = 0$$
$$+ 40 + 40I - 140 + 20 + 20I = 0$$

Combining like terms:

$$+ 60I - 80 = 0$$
$$60I = 80$$
$$I = 1.33 \text{ amps}$$

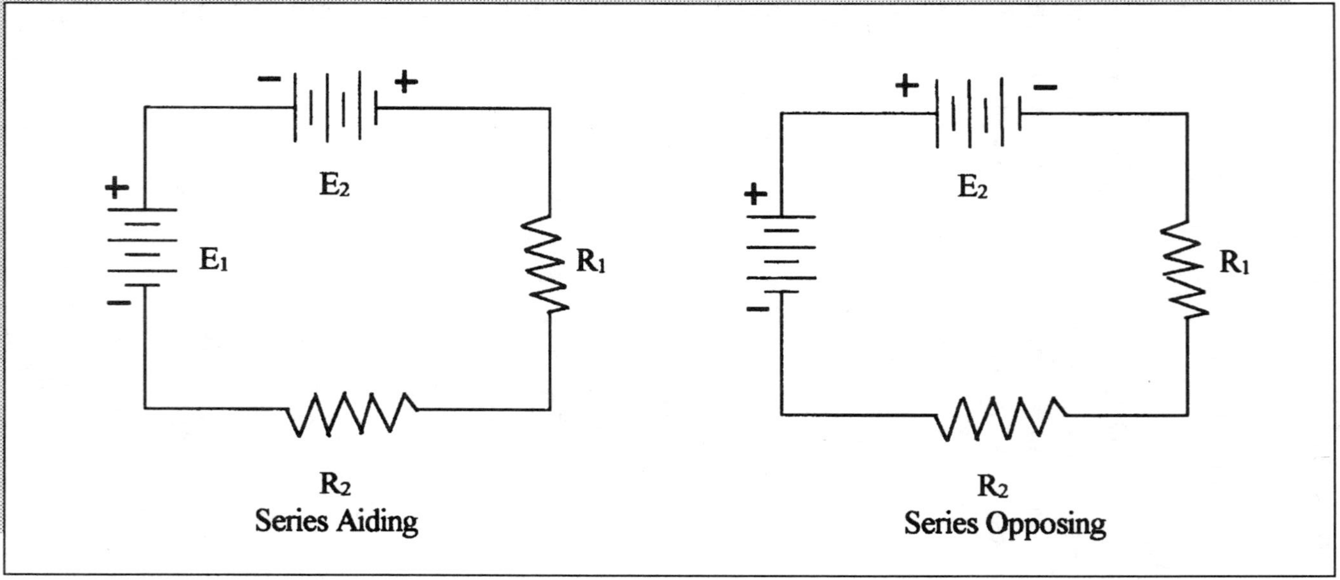

Figure 3.20
Aiding and opposing sources.

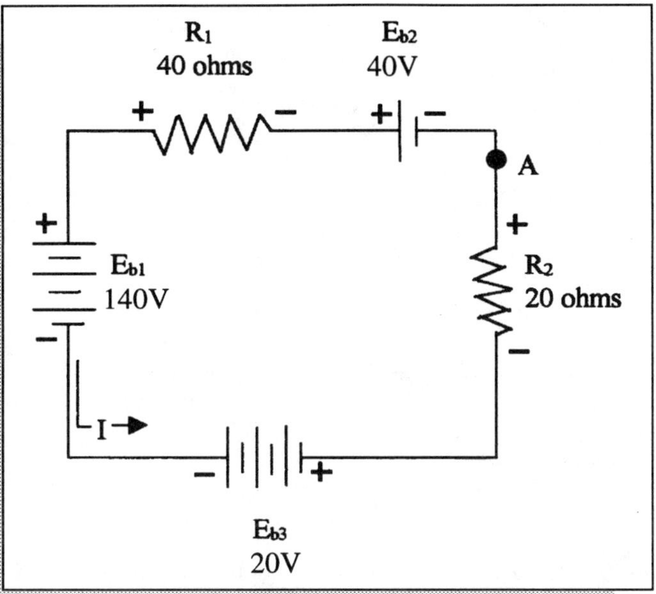

Figure 3.21
Solving for circuit current in a multiple source circuit.

3.6.11 GROUND

The term *ground* is used to denote a common electrical point of zero potential. The reference point of a circuit is always considered to be at zero potential, because the earth (ground) is said to be at a zero potential. In Figure 3.22, point A is the zero reference or ground and is symbolized as such. Point C is 60 volts positive, and point B is 20 volts positive in respect to ground.

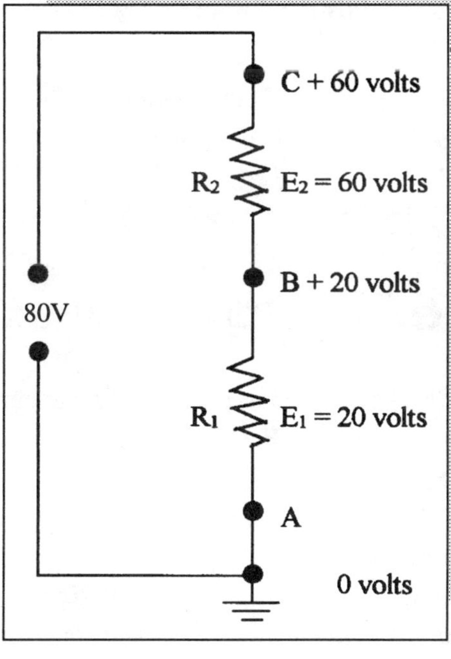

Figure 3.22
Use of ground symbols.

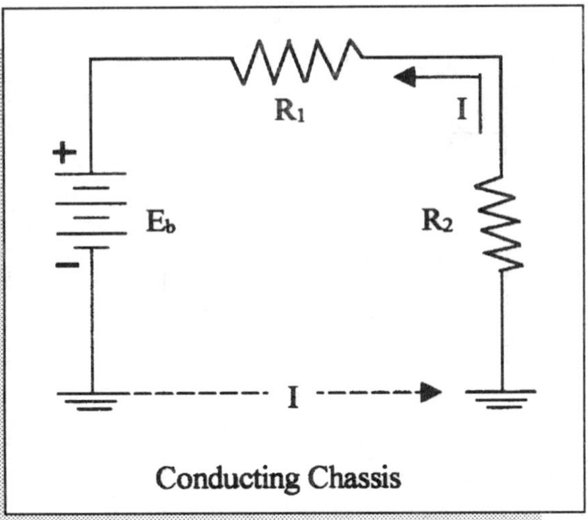

Figure 3.23
Ground used as a conductor.

The common ground for much electrical/electronic equipment is the metal chassis. The value of ground is noted when considering its contribution to economy, simplification of schematics, and ease of measurement. When completing each electrical circuit, common points of a circuit at zero potential are connected directly to the metal chassis, thereby eliminating a large amount of connecting wire. An example of a grounded circuit is illustrated in Figure 3.23.

IMPORTANT **Note:** *Most voltage measurements used to check proper circuit operation in electronic equipment are taken in respect to ground. One meter lead is attached to ground, and the other meter lead is moved to various test points.*

3.6.12 OPEN AND SHORT CIRCUITS

A circuit is *open* if a break in the circuit does not provide a complete path for current. Figure 3.24 shows an open circuit, because the fuse is blown.

To protect a circuit, a fuse is placed directly into the circuit (see Figure 3.25). A fuse will open the circuit whenever a dangerously large current starts to flow. A fuse will permit currents smaller than the fuse value to flow but will melt and therefore break or open the circuit if a

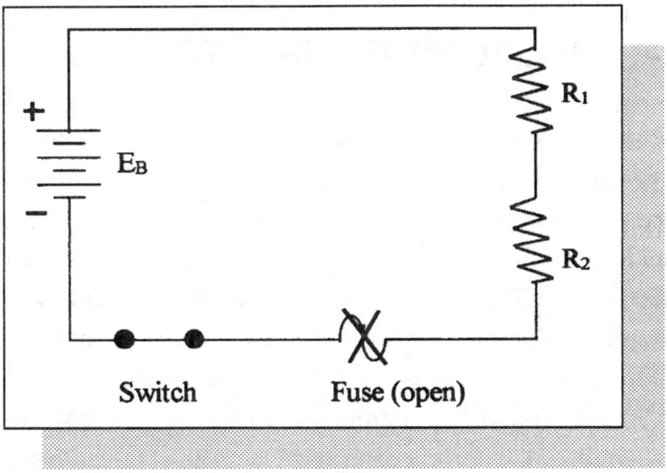

Figure 3.24
Open circuit—fuse blown.

larger current flows. A dangerously large current will flow when a "short circuit" occurs. A short circuit is usually caused by an accidental connection between two points in a circuit that offers very little resistance and passes an abnormal amount of current (see Figure 3.25). A short circuit often occurs as a result of improper wiring or broken insulation.

3.7 PARALLEL D-C CIRCUITS

The principles we applied to solving simple series circuit calculations for determining the reactions of such quantities as voltage, current, and resistance also can be used in parallel and series-parallel circuits.

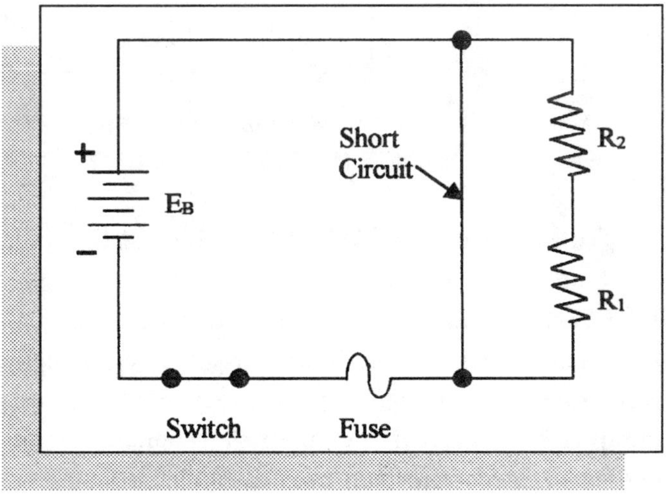

Figure 3.25
Short circuit.

3.7.1 PARALLEL CIRCUIT CHARACTERISTICS

A *parallel circuit* is defined as one having two or more components connected across the same voltage source (see Figure 3.26). Recall that in a series circuit there is only one path for current flow. As additional loads (resistors, etc.) are added to the circuit, the total resistance increases, and the total current decreases. This is *not the case* in a **parallel** circuit. In a parallel circuit, each load (or branch) is connected directly across the voltage source. In Figure 3.26, commencing at the voltage source (E_b) and tracing counterclockwise around the circuit, two complete and separate paths can be identified in which current can flow. One path is traced from the source through resistance R_1 and back to the source; the other, from the source through resistance R_2 and back to the source.

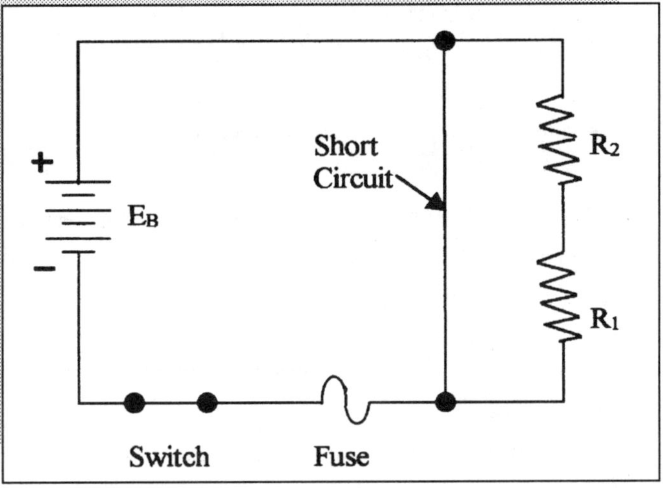

Figure 3.26 *Basic parallel circuit.*

3.7.1.1 VOLTAGE IN PARALLEL CIRCUITS

Recall that, in a series circuit, the source voltage divides proportionately across each resistor in the circuit. In a parallel circuit (see Figure 3.26), the same voltage is present across all the resistors of a parallel group. This voltage is equal to the applied voltage (E_b) and can be expressed in equation form as:

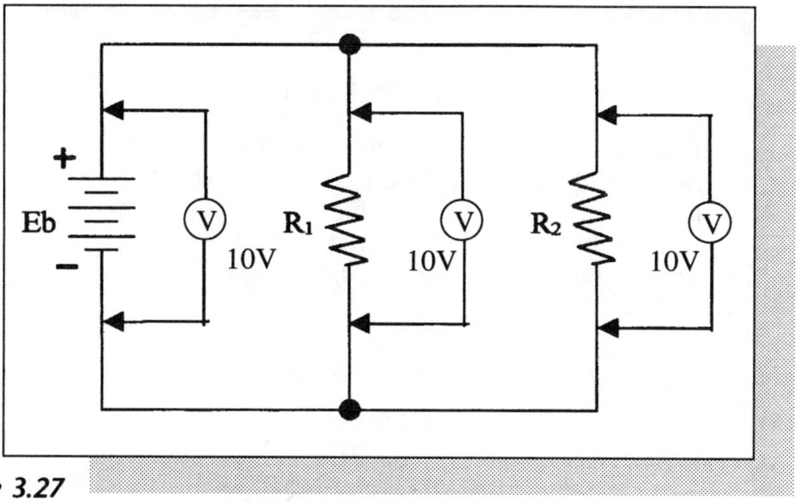

Figure 3.27
Voltage comparison in a parallel circuit.

$$E_b = E_{R1} = E_{R2} = E_{Rn} \tag{3.13}$$

We can verify Equation (3.13) by taking voltage measurements across the resistors of a parallel circuit, as illustrated in Figure 3.27. Notice that each voltmeter indicates the same amount of voltage; that is, the voltage across each resistor is the same as the applied voltage.

IMPORTANT

Key Point: *In a parallel circuit, the voltage remains the same throughout the circuit.*

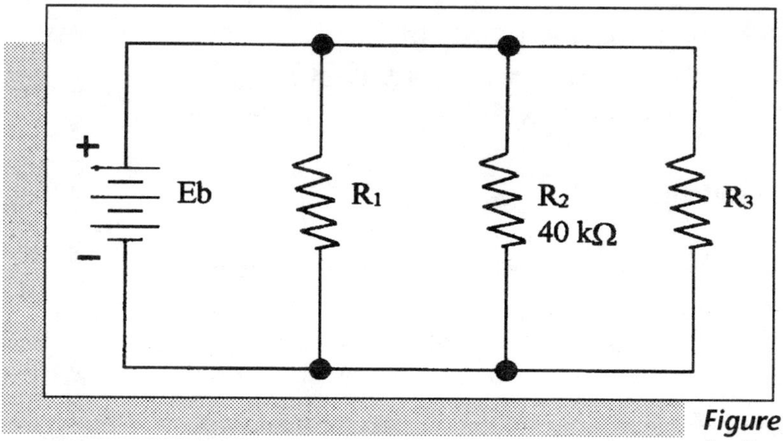

Figure 3.28
For Example 3.19.

EXAMPLE 3.19

Problem: Assume that the current through a resistor of a parallel circuit is known to be 4.0 milliamperes (ma), and the value of the resistor is 40,000 ohms. Determine the potential (voltage) across the resistor. The circuit is shown in Figure 3.28.

Solution:

Given:

$$R_2 = 40\ \text{k}\Omega$$
$$I_{R2} = 4.0\ \text{ma}$$

Find:

$$E_{R2} = ?$$
$$E_b = ?$$

Select proper equation

$$E = IR \qquad (3.3)$$

Substitute known values:

$$E_{R2} = I_{R2} \times R_2$$
$$E_{R2} = 4.0\ \text{ma} \times 40{,}000\ \text{ohms}$$
(use power of tens)
$$E_{R2} = (4.0 \times 10^{-3}) \times (40 \times 10^{3})$$
$$E_{R2} = 4.0 \times 40$$

Resultant:

$$E_{R2} = 160\ \text{V}$$

Therefore:

$$E_b = 160\ \text{V}$$

3.7.1.2 CURRENT IN PARALLEL CIRCUITS

IMPORTANT

***Important Point:** Ohm's Law states, "The current in a circuit . . . is inversely proportional to the circuit resistance." This fact, important as a basic building block of electrical theory, obviously, is also important in the following explanation of current flow in parallel circuits.*

In a series circuit, a single current flows. Its value is determined in part by the total resistance of the circuit. However, the source current in a parallel circuit divides among the available paths in relation to the value of the resistors in the circuit. Ohm's Law remains unchanged. For a given voltage, current varies inversely with resistance.

The behavior of current in a parallel circuit is best illustrated by example. The example we use is Figure 3.29. The resistors R_1, R_2, and R_3 are in parallel with each other and with the battery. Each parallel path is then a branch with its own individual current. When the total current I_T leaves the voltage source E, part I_1 of the current I_T will flow through R_1, part I_2 will flow through R_2, and the remainder I_3 through R_3. The branch current I_1, I_2, and I_3 can be different. However, if a voltmeter (used for measuring the voltage of a circuit) is connected across R_1, R_2, and R_3, the respective voltages E_1, E_2, and E_3 will be equal. Therefore,

$$E = E_1 = E_2 = E_3 \tag{3.14}$$

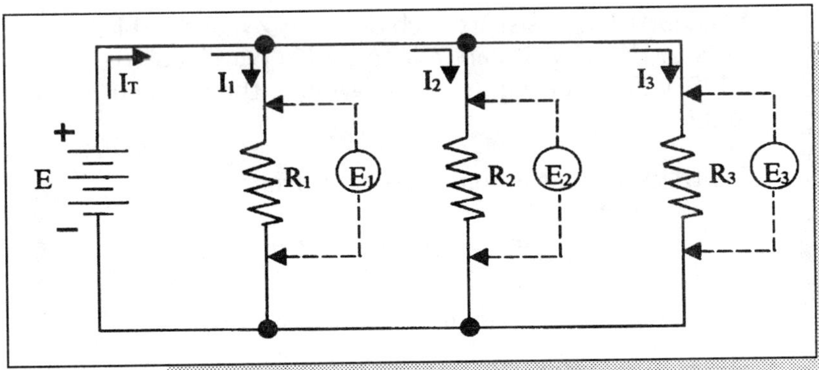

Figure 3.29
Parallel circuit.

The total current I_T is equal to the sum of all branch currents.

$$I_T = I_1 = I_2 = I_3 \tag{3.15}$$

This formula applies for any number of parallel branches whether the resistances are equal or unequal.

By Ohm's Law, each branch current equals the applied voltage divided by the resistance between the two points where the voltage is applied. Hence (Figure 3.29), for each branch, we have the following equations:

$$\begin{aligned} \text{Branch 1:} \quad I_1 &= \frac{E_1}{R_1} = \frac{E}{R_1} \\ \text{Branch 2:} \quad I_2 &= \frac{E_2}{R_2} = \frac{E}{R_2} \\ \text{Branch 3:} \quad I_3 &= \frac{E_3}{R_3} = \frac{V}{R_3} \end{aligned} \tag{3.16}$$

With the same applied voltage, any branch that has less resistance allows more current through it than a branch with higher resistance.

EXAMPLE 3.20

Problem: Two resistors each drawing 2 A and a third resistor drawing 1 A are connected in parallel across a 100-V line (see Figure 3.30). What is the total current?

Solution: The formula for total current is

$$\begin{aligned} I_T &= I_1 + I_2 + I_3 \\ &= 2 + 2 + 1 \\ &= 5 \text{ A} \end{aligned} \tag{3.15}$$

The total current is 5 A.

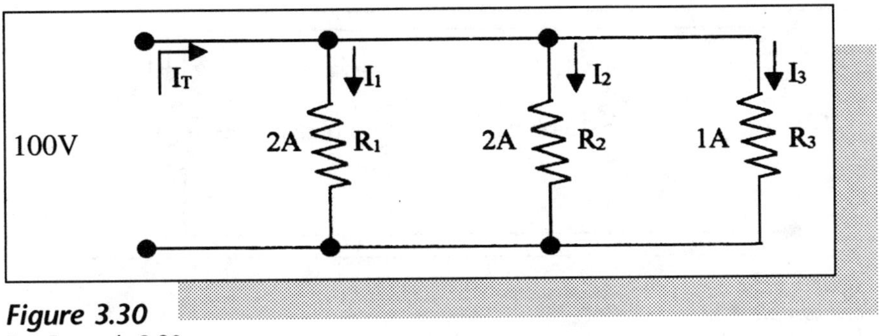

Figure 3.30
For Example 3.20.

EXAMPLE 3.21

Problem: Two branches, R_1 and R_2, across a 100-V power line draw a total line current of 20 A (Figure 3.31). Branch R_1 takes 10 A. What is the current I_2 in branch R_2?

Solution: Starting with Equation (3.15), transpose to find I_2 and then substitute given values.

$$I_T = I_1 + I_2$$
$$I_2 = I_T - I_1$$
$$= 20 - 10 = 10 \text{ A}$$

The current in branch R_2 is 10 A.

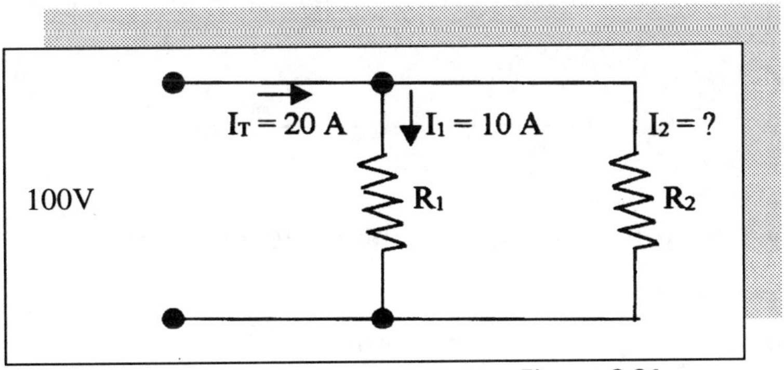

Figure 3.31
For Example 3.21.

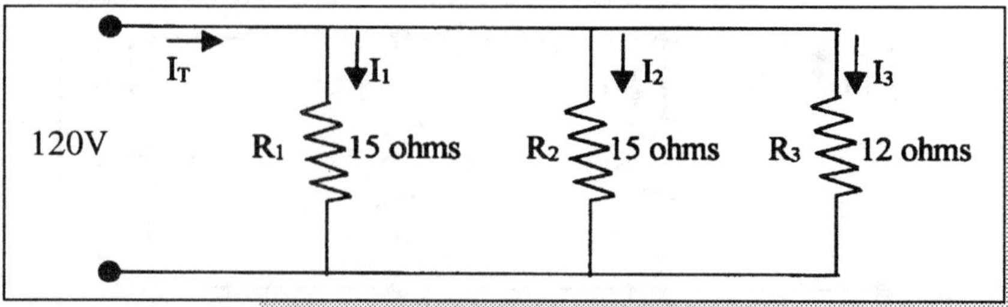

Figure 3.32
For Example 3.22.

EXAMPLE 3.22

Problem: A parallel circuit consists of two 15-ohm and one 12-ohm resistors across a 120-V line (see Figure 3.32). What current will flow in each branch of the circuit, and what is the total current drawn by all the resistors?

Solution: There is 120-V potential across each resistor. Using Equation (3.16), apply Ohm's Law to each resistor.

$$I_1 = \frac{V}{R_1} = \frac{120}{15} = 8 \text{ A}$$

$$I_2 = \frac{V}{R_2} = \frac{120}{15} = 8 \text{ A}$$

$$I_3 = \frac{V}{R_3} = \frac{120}{12} = 10 \text{ A}$$

Now find total current, using Equation (3.16).

$$I_T = I_1 + I_2 + I_3$$
$$= 8 + 8 + 10 = 26 \text{ A}$$

3.7.1.2.1 PARALLEL CIRCUITS AND KIRCHHOFF'S CURRENT LAW

The division of current in a parallel network follows a definite pattern. This pattern is described by *Kirchhoff's Current Law,* which is stated as follows:

The algebraic sum of the currents entering and leaving any junction of conductors is equal to zero. This can be stated mathematically as

$$I_a + I_b + \ldots\ldots + I_n = 0 \qquad (3.17)$$

where I_a, I_b + etc. are the currents entering and leaving the junction. Currents entering the junction are assumed to be positive, and currents leaving the junction are considered negative. When solving a problem using Equation (3.17), the currents must be placed into the equation with the proper polarity.

EXAMPLE 3.23

Problem: Solve for the value of I_3 in Figure 3.33.

Solution: First, the currents are given proper signs.

I_1 = + 10 amps
I_2 = − 3 amps
I_3 = ? amps
I_4 = − 5 amps

Then, place these currents into Equation (3.18) with the proper signs as follows:

Basic equation: $I_a + I_b + \ldots\ldots I_n = 0 \qquad (3.17)$

Substitution: $I_1 + I_2 + I_3 + I_4 = 0$

$(+10) + (-3) + (I_3) + (-5) = 0$

Combining like terms:

$I_3 = 2 = 0$

$I_3 = -2$ amps

thus, I_3 has a value of 2 amps, and the negative sign shows it to be a current leaving the junction.

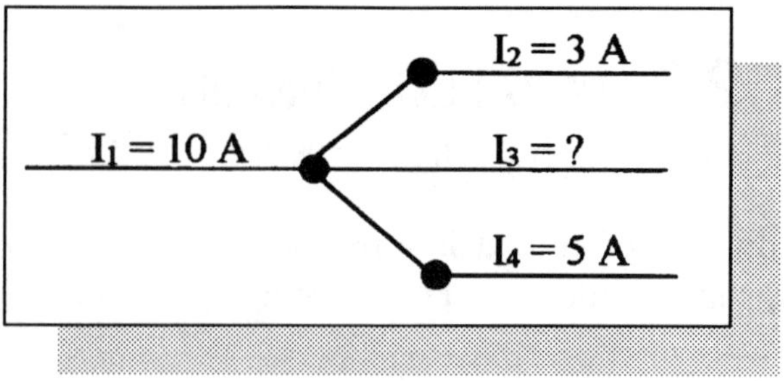

Figure 3.33
For Example 3.23.

3.7.1.3 PARALLEL RESISTANCE

Unlike series circuits where total resistance (R_T) is the sum of the individual resistances, in a parallel circuit, the total resistance is *not* the sum of the individual resistances.

In a parallel circuit, we can use Ohm's Law to find total resistance. We use the equation

$$R = \frac{E}{I} \tag{3.2}$$

or

$$R_T = \frac{E_S}{I_T}$$

R_T is the total resistance of all the parallel branches across the voltage source E_S, and I_T is the sum of all the branch currents.

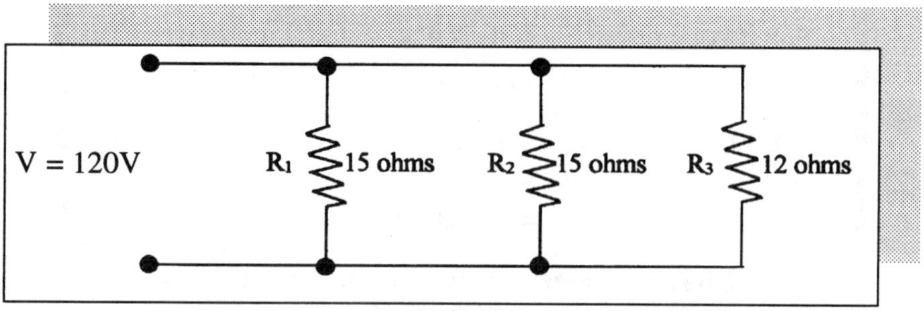

Figure 3.34
For Example 3.24.

EXAMPLE 3.24

Problem: What is the total resistance of the circuit shown in Figure 3.34?

Given: $E_S = 120$ V
$I_T = 26$ A

Solution: In Figure 3.34, the line voltage is 120 V, and the total line current is 26 A. Therefore,

$$R_T = \frac{E}{I_T} = \frac{120}{26} = 4.62 \text{ ohms}$$

IMPORTANT

Important Point: *Notice that* R_T *is smaller than any of the three resistances in Figure 3.34. This fact may surprise you—it may seem strange that the total circuit resistance is* less *than that of the smallest resistor* (R_3 *– 12 ohms). However, if we refer back to the water analogy we have used previously, it makes sense. Consider water pressure and water pipes—and assume there is some way to keep the water pressure constant. A small pipe offers more resistance to flow of water than a larger pipe; but if we add another pipe in parallel, one of even smaller diameter, the total resistance to water flow is* ***decreased.*** *In an electrical circuit, even a larger resistor in another parallel branch provides an* ***additional path for current flow,*** *so the* ***total*** *resistance is less. Remember, if we add one more branch to a parallel circuit, the total resistance decreases, and the total current increases.*

Back to Example 3.24 and Figure 3.34. What we essentially demonstrated in working this particular problem is that the total load connected to the 120-V line is the same as the single equivalent resistance of 4.62 Ω connected across the line. (It is probably more accurate to call this total resistance the "equivalent resistance," but by convention R_T, or total resistance, is used—but they are often used interchangeably, too.)

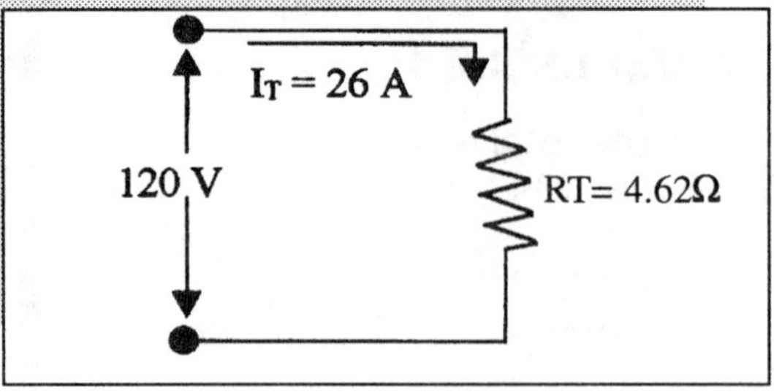

Figure 3.35
Equivalent circuit to that of Figure 3.34.

We illustrate the "equivalent resistance" in the equivalent circuit shown in Figure 3.35.

There are other methods used to determine the equivalent resistance of parallel circuits. The most appropriate method for a particular circuit depends on the number and value of the resistors. For example, consider the parallel circuit shown in Figure 3.36.

For this circuit, the following simple equation is used:

$$R_{eq} = \frac{R}{N} \qquad (3.18)$$

where

R_{eq} = equivalent parallel resistance
R = ohmic value of one resistor
N = number of resistors

Thus,

$$R_{eq} = \frac{10 \text{ ohms}}{2}$$

$$R_{eq} = 5 \text{ ohms}$$

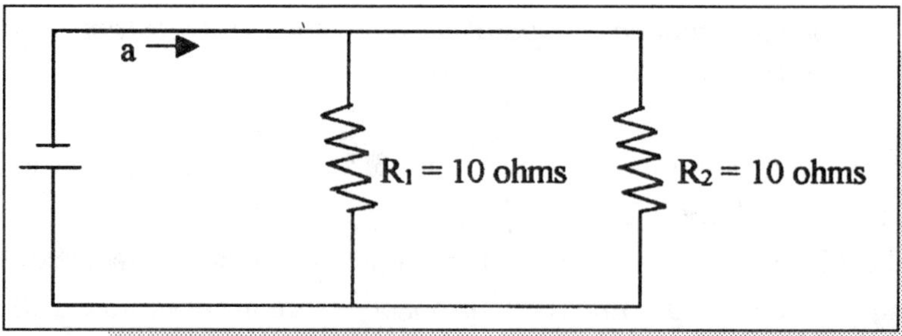

Figure 3.36
Two equal resistors connected in parallel.

IMPORTANT

Note: *Equation (3.18) is valid for any number of equal value parallel resistors.*

IMPORTANT

Key Point: *When two equal value resistors are connected in parallel, they present a total resistance equivalent to a single resistor of one-half the value of either of the original resistors.*

EXAMPLE 3.25

Problem: Five 50-ohm resistors are connected in parallel. What is the equivalent circuit resistance?

Solution: Using Equation (3.18)

$$R_{eq} = \frac{R}{N} = \frac{50}{5} = 10 \text{ ohms}$$

What about parallel circuits containing resistance of unequal value? How is equivalent resistance determined?

Example 3.26 demonstrates how this is accomplished.

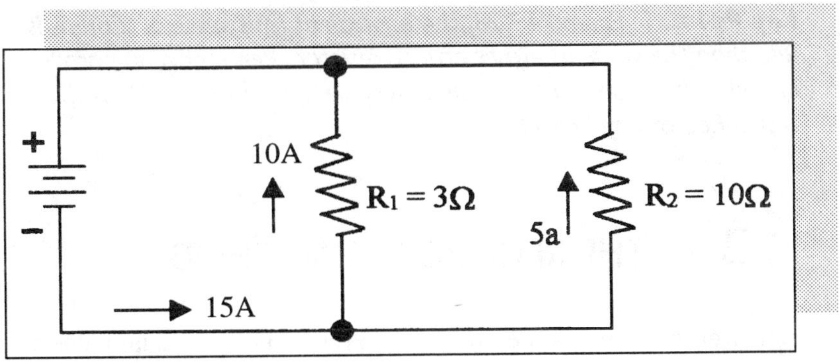

Figure 3.37
For Example 3.26.

EXAMPLE 3.26

Problem: Refer to Figure 3.37.

Solution:

Given:
$$R_1 = 3\ \Omega$$
$$R_2 = 6\ \Omega$$
$$E_a = 30\ V$$

Known:
$$I_1 = 10 \text{ amps}$$
$$I_2 = 5 \text{ amps}$$
$$I_t = 15 \text{ amps}$$

Determine:
$$R_{eq} = ?$$

$$R_{eq} = \frac{E_a}{I_t}$$

$$R_{eq} = \frac{30}{15} = 2 \text{ ohms}$$

IMPORTANT

Key Point: *In Example 3.28, the equivalent resistance of 2 ohms is less than the value of either branch resistor. Remember, in parallel circuits, the equivalent resistance will always be smaller than the resistance of any branch.*

3.7.1.3.1 THE RECIPROCAL METHOD

When circuits are encountered in which resistors of unequal value are connected in parallel, the equivalent resistance may be computed by using the *reciprocal method*.

IMPORTANT

Note: *A* reciprocal *is an inverted fraction; the reciprocal of the fraction 3/4, for example is 4/3. We consider a whole number to be a fraction with 1 as the denominator, so the reciprocal of a whole number is that number divided into 1. For example, the reciprocal of* R_t *is* $1/R_t$.

The equivalent resistance in parallel is given by the formula

$$\frac{1}{R_T} = \frac{1}{R_1} + \frac{1}{R_2} + \frac{1}{R_3} + \ldots + \frac{1}{R_n} \quad (3.19)$$

where R_T *is the total resistance in parallel and* R_1, R_2, R_3, *and* R_n *are the branch resistances.*

EXAMPLE 3.27

Problem: Find the total resistance of a 2-ohm, a 4-ohm, and an 8-ohm resistor in parallel (Figure 3.38).

Solution: Write the formula for three resistors in parallel.

$$\frac{1}{R_T} = \frac{1}{R_1} + \frac{1}{R_2} + \frac{1}{R_3} \quad (3.19)$$

Substituting the resistance values.

$$\frac{1}{R_T} = \frac{1}{2} + \frac{1}{4} + \frac{1}{8}$$

Add fractions.

$$\frac{1}{R_T} = \frac{4}{8} + \frac{2}{8} + \frac{1}{8} = \frac{7}{8}$$

Invert both sides of the equation to solve for R_T.

$$R_T = \frac{8}{7} = 1.14\ \Omega$$

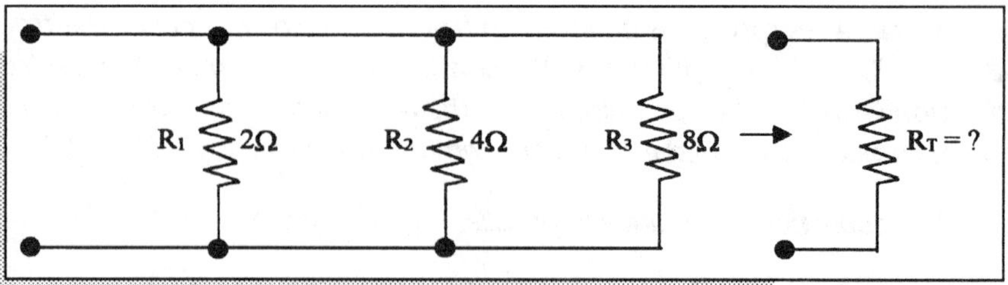

Figure 3.38
For Example 3.27.

IMPORTANT **Note:** *When resistances are connected in parallel, the total resistance is always* **less** *than the smallest resistance of any single branch.*

3.7.1.3.2 PRODUCT OVER THE SUM METHOD

When any two unequal resistors are in parallel, it is often easier to calculate the total resistance by multiplying the two resistances and then dividing the product by the sum of the resistances.

$$R_T = \frac{R_1 \times R_2}{R_1 + R_2} \qquad (3.20)$$

where R_T is the total resistance in parallel and R_1 and R_2 are the two resistors in parallel.

EXAMPLE 3.28

Problem: What is the equivalent resistance of a 20-ohm and a 30-ohm resistor connected in parallel?

Solution:

Given:

$$R_1 = 20$$
$$R_2 = 30$$

$$R_T = \frac{R_1 \times R_2}{R_1 + R_2} \qquad (3.20)$$

$$R_T = \frac{20 \times 30}{20 + 30}$$

$$R_T = 12 \text{ ohms}$$

3.7.1.3.3 REDUCTION TO AN EQUIVALENT CIRCUIT

In the study of basic electricity, it is often necessary to resolve a complex circuit into a simpler form. Any complex circuit consisting of resistances can be reduced to a basic equivalent circuit containing the source and total resistance. This process is called *reduction to an equivalent circuit.* An example of circuit reduction was demonstrated in Example 3.28 and is illustrated in Figure 3.39.

The circuit shown in Figure 3.39(a) is reduced to the simple circuit shown in (b).

3.7.1.4 POWER IN PARALLEL CIRCUITS

As in the series circuit, the total *power* consumed in a parallel circuit is equal to the sum of the power consumed in the individual resistors.

IMPORTANT **Note:** *Because power dissipation in resistors consists of a heat loss, power dissipations are additive regardless of how the resistors are connected in the circuit.*

$$P_T = P_1 + P_2 + P_3 + \ldots + P_n \qquad (3.21)$$

where P_T is the total power and P_1, P_2, P_3, and P_n are the branch powers.

Total power can also be calculated by the equation

$$P_T = EI_T \qquad (3.22)$$

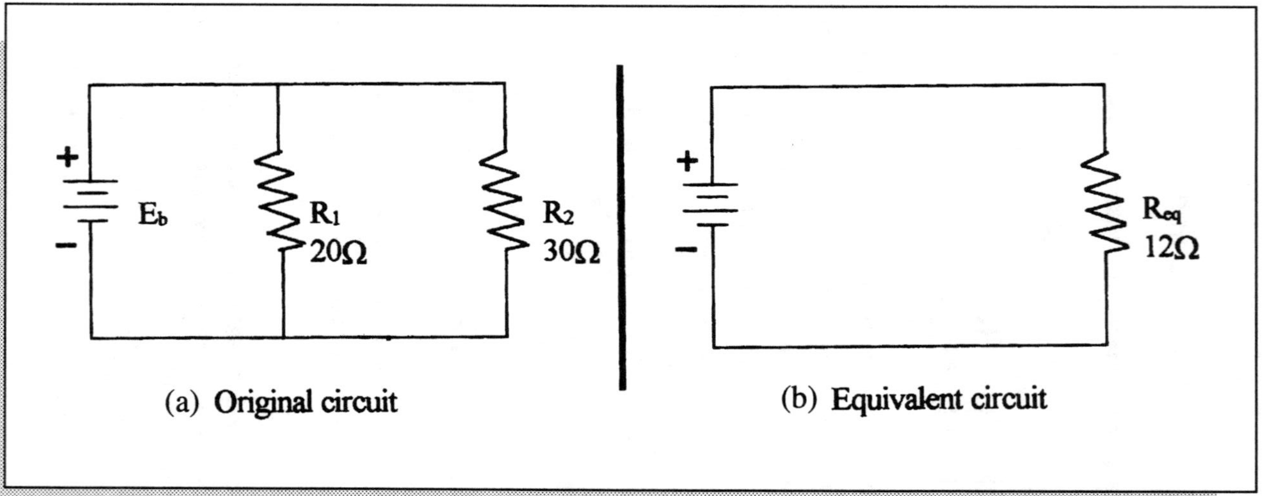

Figure 3.39
Parallel circuit with equivalent circuit.

where P_T is the total power, E is the voltage source across all parallel branches, and I_T is the total current.

The power dissipated in each branch is equal to EI and equal to V^2/R.

IMPORTANT

Note: *In both parallel and series arrangements, the sum of the individual values of power dissipated in the circuit equals the total power generated by the source. The circuit arrangements cannot change the fact that all power in the circuit comes from the source.*

IMPORTANT

Note: *The power dissipated in the branch circuits in Figure 3.40 is determined in the same manner as the power dissipated by individual resistors in a series circuit. The total power (P_t) is then obtained by summing up the powers dissipated in the branch resistors using Equation (3.21).*

Because, in the example shown in Figure 3.40, the total current is known, we could determine the total power by the following method:

$$P_T = E_b \times I_T$$

$$P_T = 50\text{ V} \times 8\text{A}$$

$$P_T = 400\text{ watts}$$

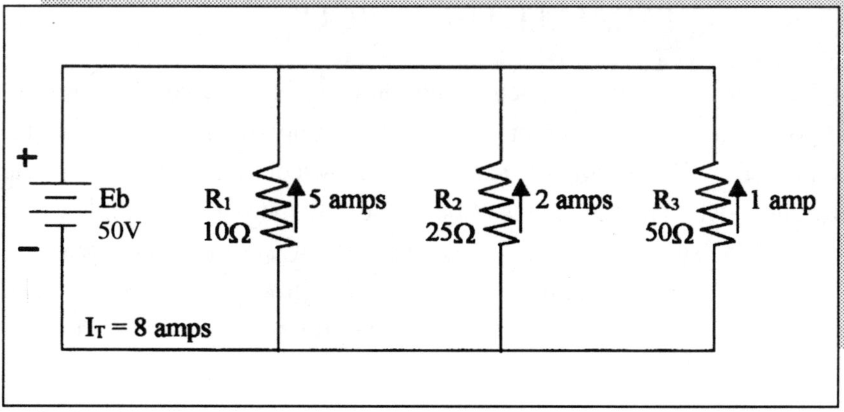

Figure 3.40
For Example 3.29.

EXAMPLE 3.29

Problem: Find the total power consumed by the circuit in Figure 3.40.

Solution:

$P_{R1} = E_b \times I_{R1}$

$P_{R1} = 50 \times 5$

$P_{R1} = 250$ watts

$P_{R2} = E_b \times I_{R2}$

$P_{R2} = 50 \times 2$

$P_{R2} = 100$ watts

$P_{R3} = E_b \times I_{R3}$

$P_{R3} = 50 \times 1$

$P_{R3} = 50$ watts

$P_T = P_1 + P_2 + P_3$

$P_T = 250 + 100 + 50$

$P_T = 400$ watts

3.7.1.5 RULES FOR SOLVING PARALLEL D-C CIRCUITS

Problems involving the determination of resistance, voltage, current, and power in a parallel circuit are solved as simply as in a series circuit. The procedure is basically the same—(1) draw a circuit diagram, (2) state the values given and the values to be found, (3) state the applicable equations, and (4) substitute the given values and solve for the unknown.

Along with following the problem-solving procedure above, it is also important to remember and apply the rules for solving parallel d-c circuits:

- The same voltage exists across each branch of a parallel circuit and is equal to the source voltage.

- The current through a branch of a parallel network is inversely proportional to the amount of resistance of the branch.
- The total current of a parallel circuit is equal to the sum of the currents of the individual branches of the circuit.
- The total resistance of a parallel circuit is equal to the reciprocal of the sum of the reciprocals of the individual resistances of the circuit.
- The total power consumed in a parallel circuit is equal to the sum of the power consumption of the individual resistances.

3.8 SERIES-PARALLEL CIRCUITS

To this point, we have discussed series and parallel d-c circuits. However, the maintenance operator will seldom encounter a circuit that consists solely of either type of circuit. Most circuits consist of both series and parallel elements. A circuit of this type is referred to as a *series-parallel circuit,* or a combination circuit. The solution of a series-parallel (combination) circuit is simply a matter of application of the laws and rules discussed prior to this point.

3.8.1 SOLVING A SERIES-PARALLEL CIRCUIT

At least three resistors are required to form a series-parallel circuit. An example of a series-parallel circuit is shown in Figure 3.41, where two parallel resistors, R_2 and R_3, are connected in series with the resistor R_1 and the voltage source E. In a circuit of this type, the current I_T divides after it flows through R_1, and part flows through R_2 and part flows through R_3. Then, the current joins at the junction of the two resistors and flows back to the positive terminal of the voltage source (E) and through the voltage source to the positive terminal.

In solving for values in a series-parallel circuit (current, voltage, and resistance), follow the rules that apply to a series circuit for the series part

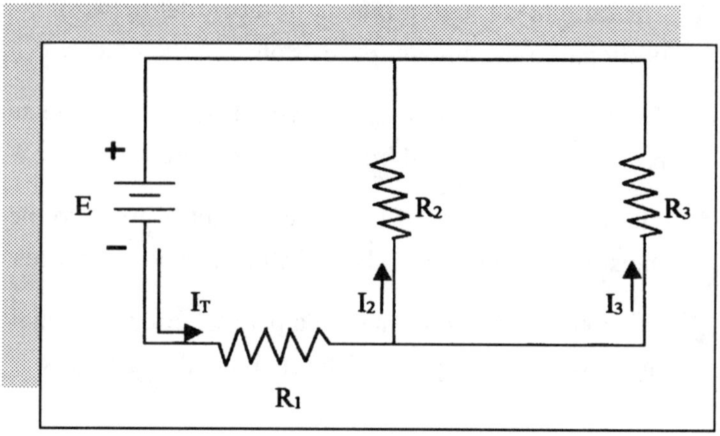

Figure 3.41
A series-parallel circuit.

of the circuit, and follow the rules that apply to a parallel circuit for the parallel part of the circuit. Solving series-parallel circuits is simplified if all parallel and series groups are first reduced to single equivalent resistances and the circuits redrawn in simplified form. Recall, the redrawn circuit is called an **equivalent circuit.** The procedure for developing an equivalent circuit is shown in Figure 3.42.

IMPORTANT ***Note:*** *There are no general formulas for the solution of series-parallel circuits because there are so many different forms of these circuits.*

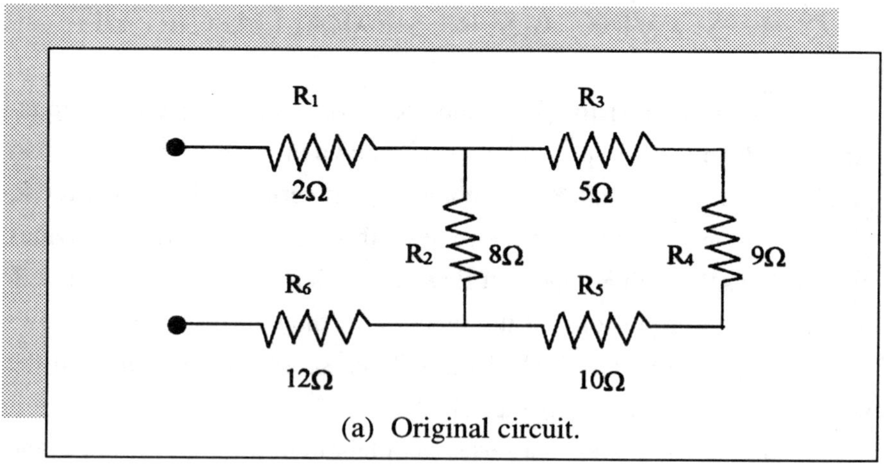

Figure 3.42 *Developing an equivalent circuit.*

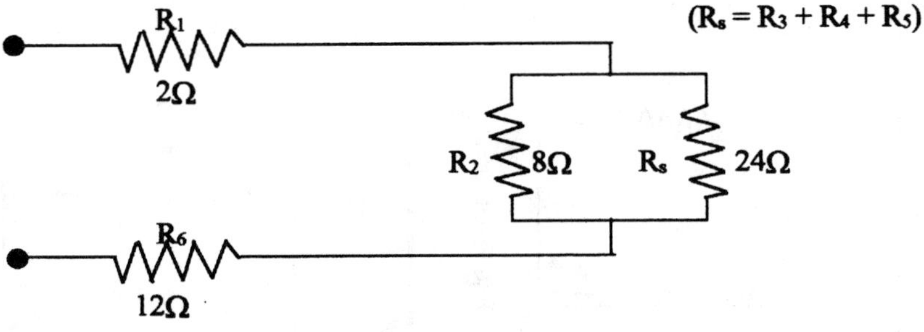

(b) First simplification.

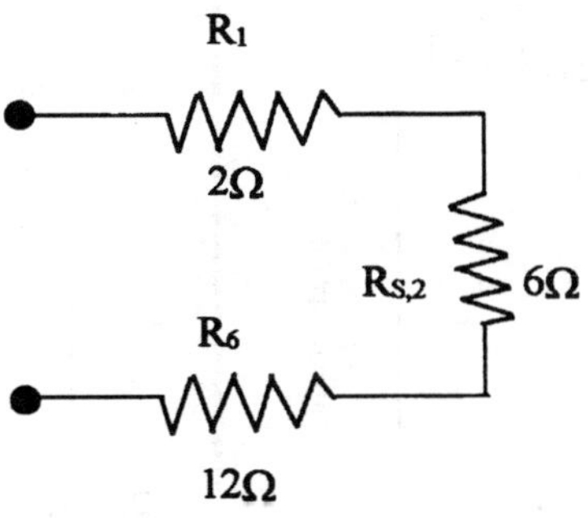

(c) Second simplification.

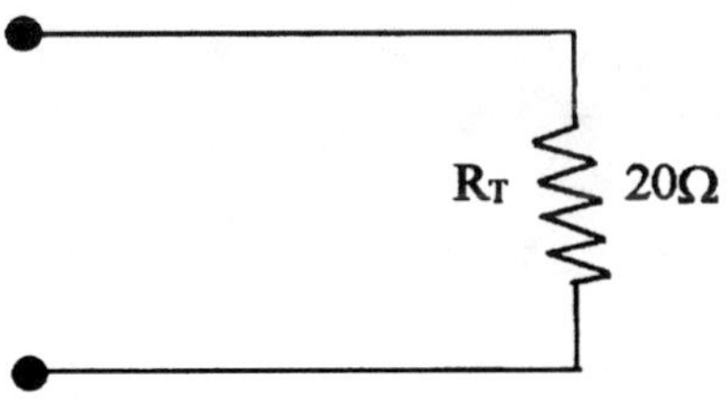

(d) Equivalent resistance

Figure 3.42 (continued)
Developing an equivalent circuit.

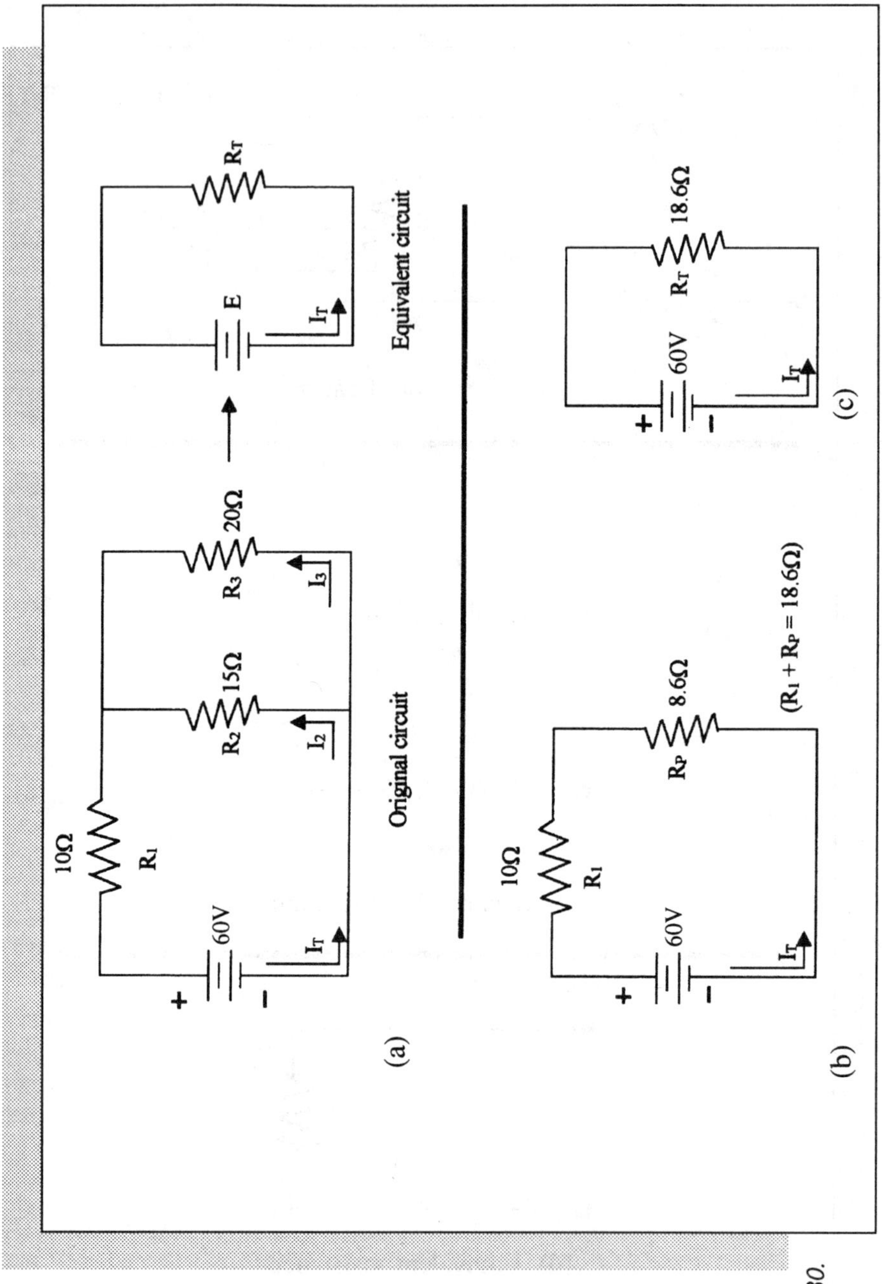

***Figure* 3.43**
For Example 3.30.

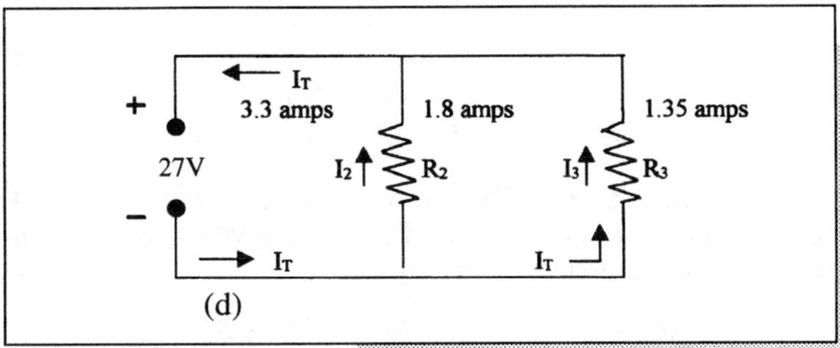

Figure 3.43 (continued)
For Example 3.30.

EXAMPLE 3.30

Problem:

Find the total resistance (R_T), total circuit current (I_T), and branch currents of the circuit shown in Figure 3.43 (a).

Solution:

Step 1: Find the equivalent resistance of the parallel branch.

$$R_p = \frac{R_2R_3}{R_2 + R_3} = \frac{(15)(20)}{15 + 20} = \frac{300}{35} = 8.6\ \Omega$$

The equivalent circuit reduces to a series circuit [Figure 3.43(b)].

Step 2: Find the resistance of the equivalent series circuit.

$$R_T = R_1 + R_P = 10 + 8.6 = 18.6\ \Omega$$

The equivalent circuit reduces to a single voltage source and a single resistance [Figure 3.43(c)].

Step 3: Find I_T. (I_T is the actual current being supplied in the original series-parallel circuit.)

$$I_T = \frac{V}{R_T} = \frac{60}{18.6\ \Omega} = 3.2 \text{ amps}$$

Step 4: Find I_2 and I_3. The voltage across R_2 and R_3 is equal to the applied voltage (E) less the voltage drop across R_1 [Figure 3.48(d)].

$$V_2 = V_3 = V - I_T R_1 = 60 - (3.2 \times 10) = 28 \text{ volts}$$

Then,

$$I_2 = \frac{V_2}{R_2} = \frac{28}{15} = 1.9 \text{ amp}$$

$$I_3 = \frac{V_3}{R_3} = \frac{28}{20} = 1.4 \text{ amp}$$

IMPORTANT

Important Note: *The total current in the series-parallel circuit depends on the effective resistance of the parallel portion* ***and*** *on the other resistances.*

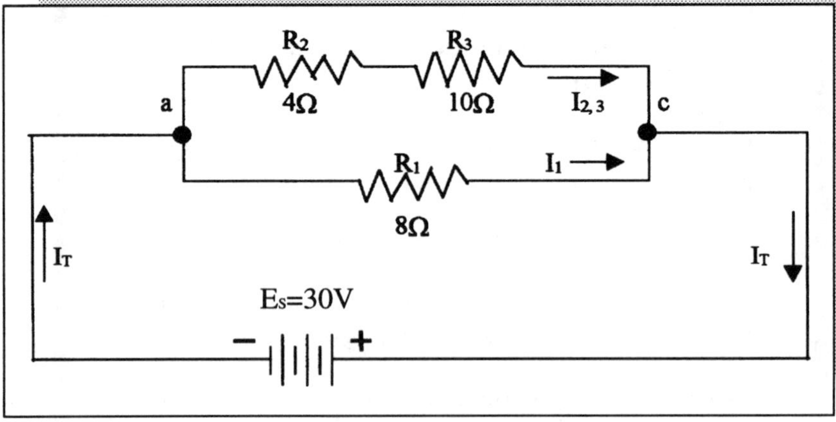

Figure 3.44
For Example 3.31.

EXAMPLE 3.31

Problem: Answer the following questions related to Figure 3.44.

Question 1:

Figure 3.44 shows a circuit in which two resistors in series (R_2 and R_3) form one branch of a parallel circuit. The total current (I_T) flows into the parallel circuit, splitting into two branches at point a. At what point do the branch currents rejoin to form I_T?

Answer:

At point c.

Question 2:

The source voltage E_S (30 V) is dropped between points a and c. The largest voltage drop is across which resistor? Why?

Answer:

R_1, because the 30 V must be divided between R_2 and R_3.

Question 3:

What is the resistance of the top branch of the circuit in Figure 3.44?

Answer:

$$14\ \Omega\ (\text{add } R_2 + R_3)$$

Question 4:

$R_T = ?$

Answer:

$$R_T = \frac{8 \times 14}{8 + 14} = \frac{112}{22} = 5.1\ \Omega$$

Question 5:

What is the current through the top branch ($I_{2,3}$)?

Answer:

$$I = \frac{E}{R} = \frac{30\ V}{14\ \Omega} = 2.14 \text{ amps}$$

Question 6:

$P_T = ?$

Answer:

$$P_T = \frac{E_{S2}}{R_t}$$

$$P_T = \frac{30\ V2}{5.1\ \Omega}$$

$$P_T = 176.5 \text{ watts}$$

Self-Test

3.1 Explain the difference between a series and a parallel circuit.

3.2 The sum of all voltages in a series circuit is equal to______________.

3.3 For any total voltage rise in a circuit, there must be an equal total ______________.

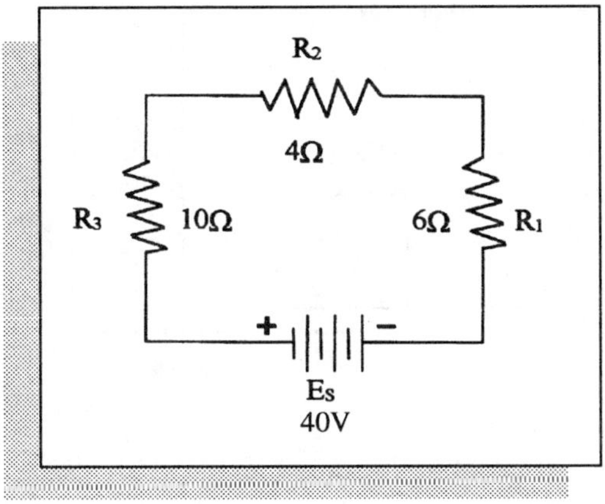

Figure 3.45
For questions 3.4–3.8.

Refer to Figure 3.45 for questions 3.4 through 3.8.

3.4 Is the direction of current flow clockwise or counterclockwise?

3.5 $I_T = ?$

3.6 E dropped across $R_1 = ?$

3.7 What is the power absorbed by R_2?

3.8 $P_T = ?$

3.9 The equivalent resistance R_T of parallel branches is ______________ than the smallest branch resistance because all the branches must take ______________ current from the source than any one branch.

3.10 A short circuit has __________ resistance, resulting in __________ current.

3.11 What is the total resistance of four 30-Ω resistors connected in series?

3.12 There is only __________ voltage across all components in parallel.

3.13 The sum of the __________ values of power dissipated in parallel resistances equals the __________ power produced by the source.

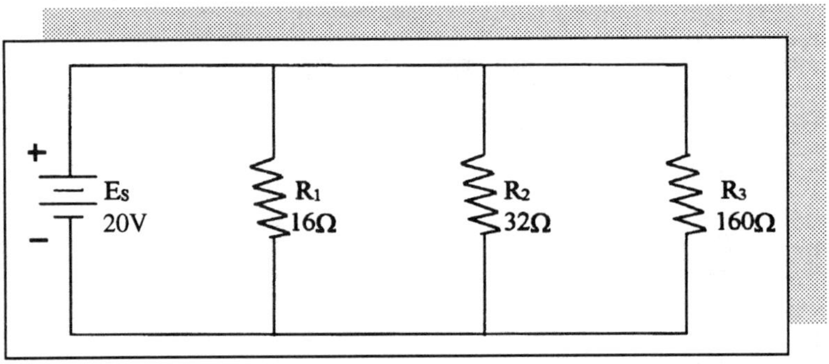

Figure 3.46
For questions 3.14–3.16.

Refer to Figure 3.46 for questions 3.14 through 3.16.

3.14 $R_T = ?$

3.15 $I_T = ?$

3.16 $P_T = ?$

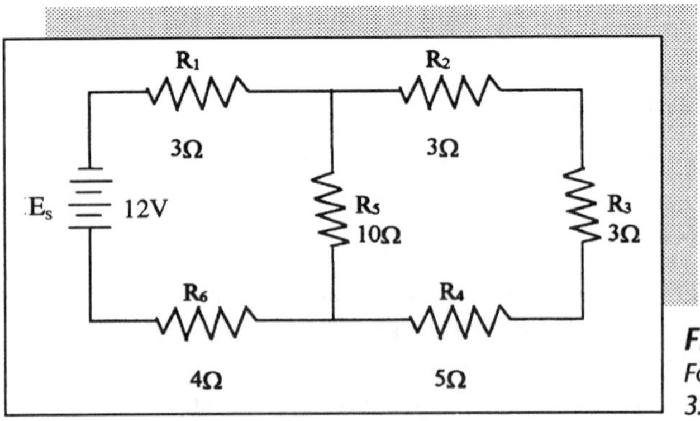

Figure 3.47
For questions 3.17–3.20.

Refer to Figure 3.47 for questions 3.17 through 3.20.

3.17 What is the total resistance of the circuit?

3.18 What is the total current I_T in the circuit?

3.19 What is the voltage across R_3?

3.20 What is the total power consumed in the circuit?

Conductors

TOPICS

Unit Size of Conductors
Resistivity
Wire Measurement
Factors Governing Selection of Wire Size
Temperature Coefficient
Conductor Insulation
Conductor Splices and Terminal Connections
Soldering Operations
Solderless Connectors
Insulation Tape

4.1 INTRODUCTION

Recall we pointed out earlier (Section 1.4) that electric current moves easily through some materials but with greater difficulty through others. Three good electrical conductors are copper, silver, and aluminum (generally, we can say that most metals are good conductors). At the present time, copper is the material of choice used in electrical conductors. Under special conditions, certain gases are also used as conductors (e.g., neon gas, mercury vapor, and sodium vapor are used in various kinds of lamps).

The function of the wire conductor is to connect a source of applied voltage to a load resistance with a minimum *IR* voltage drop in the conductor so that most of the applied voltage can produce current in the load resistance. Ideally, a conductor must have a very low resistance (e.g., a typical value for a conductor—copper—is less than 1 Ω per 10 feet).

Because all electrical circuits utilize conductors of one type or another, in this chapter, we discuss the basic features and electrical characteristics of the most common types of conductors.

Key Terms Used in This Chapter

CONDUCTOR	A material or substance that passes electrical current with ease.
INSULATOR	A material that does not normally pass electrical current.
MIL	0.001 of an inch.
MIL-FOOT	A unit of measurement for size of wire.
RESISTIVITY	The resistance in ohms of a unit conductor or a given substance.
CIRCULAR MIL	Standard unit of wire cross-sectional area used in most wire tables.
TEMPERATURE COEFFICIENT	Indicates how much resistance changes for a change in temperature.
INSULATION RESISTANCE	The resistance to current leakage through and over the surface of insulation materials.
DIELECTRIC STRENGTH	The ability of an insulator to withstand potential difference.

Moreover, because conductor splices and connections (and insulation of such connections) are also an essential part of any electric circuit, they are also discussed.

4.2 UNIT SIZE OF CONDUCTORS

A standard (or unit size) of a conductor has been established to compare the resistance and size of one conductor with another. The unit of linear measurement used is (in regard to diameter of a piece of wire) the **mil** (0.001 of an inch). A convenient unit of wire length used is the **foot.** Thus, the standard unit of size in most cases is the **mil-foot** (i.e., a wire will have unit size if it has diameter of 1 mil and a length of 1 foot). The resistance in ohms of a unit conductor or a given substance is called the **resistivity** (or specific resistance) of the substance.

As a further convenience, **gage** numbers are also used in comparing the diameter of wires. The B and S (Browne and Sharpe) gage was used in the past; now, the most commonly used gage is the **American Wire Gage** (AWG).

4.2.1 SQUARE MIL

Figure 4.1 shows a *square mil.* The square mil is a convenient unit of cross-sectional area for square or rectangular conductors. As shown in Figure 4.1, a square mil is the area of a square, the sides of which are 1 mil. To obtain the cross-sectional area in square mils of a square conductor, square one side measured in mils. To obtain the cross-sectional area in square mils of a rectangular conductor, multiply the length of one side by that of the other, each length being expressed in mils.

EXAMPLE 4.1

Problem:

Find the cross-sectional area of a large rectangular conductor 5/8 inch thick and 5 inches wide.

Solution:

The thickness may be expressed in mils as 0.625 in x 1000 mil/in = 625 mils and the width as 5 in x 1000 mil/in = 5000 mils. The cross-sectional area is 625 x 5000, or 3,125,000 square mils.

4.2.2 CIRCULAR MIL

The *circular mil* is the standard unit of wire cross-sectional area used in most wire tables. To avoid the use of decimals (because most wires used to conduct electricity may be only a small fraction of an inch), it is convenient to express these diameters in mils. For example, the diameter of a wire is expressed as 25 mils instead of 0.025 inch. A circular mil is

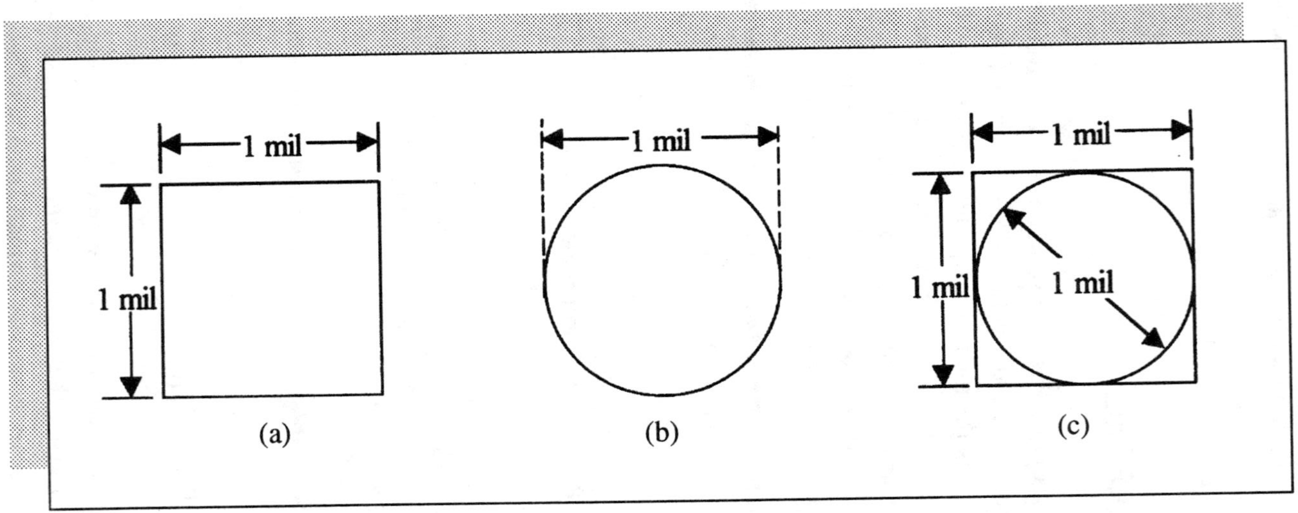

Figure 4.1
(a) Square mil (b) circular mil
(c) comparison of circular to square mil.

the area of a circle having a diameter of 1 mil, as shown in Figure 4.1(b). The area in circular mils of a round conductor is obtained by squaring the diameter measured in mils. Thus, a wire having a diameter of 25 mils has an area of 252 or 625 circular mils. By way of comparison, the basic formula for the area of a circle is

$$\mathbf{A = \pi R_2} \tag{4.1}$$

and, in this example, the area in square inches is

$$A = \pi R^2 = 3.14(0.0125)^2 =$$
$$0.00049 \text{ sq in}$$

If D is the diameter of a wire in mils, the area in square mils can be determined using

$$\mathbf{A = \pi(D/2)^2} \tag{4.2}$$

which translates to

$$= 3.14/4\ D^2$$
$$= 0.785\ D^2 \text{ sq mils}$$

Thus, a wire 1 mil in diameter has an area of

$$A = 0.785 \times 1^2 = 0.785 \text{ sq mils}$$

which is equivalent to 1 circular mil. The cross-sectional area of a wire in circular mils is therefore determined as

$$A = \frac{0.785\ D^2}{0.785} = D^2 \text{ circular mils}$$

where D is the diameter in mils. Therefore, the constant p/4 is eliminated from the calculation.

It should be noted that, in comparing square and round conductors, the circular mil is a smaller unit of area than the square mil, and, therefore, there are more circular mils than square mils in any given area. The comparison is shown in Figure 4.1(c). The area of a circular mil is equal to 0.785 of a square mil.

IMPORTANT ***Important Point:*** *To determine the circular-mil area when the square-mil area is given, divide the area in square mils by 0.785. Conversely, to determine the square-mil area when the circular-mil area is given, multiply the area in circular mils by 0.785.*

EXAMPLE 4.2

Problem: A No. 12 wire has a diameter of 80.81 mils. What is (1) its area in circular mils and (2) its area in square mils?

Solution:

(1) $A = D^2 = 80.81^2 = 6530$ circular mils

(2) $A = 0.785 \times 6530 = 5126$ square mils

EXAMPLE 4.3

Problem: A rectangular conductor is 1.5 inches wide and 0.25 inch thick. (1) What is its area in square mils? (2) What size of round conductor in circular mils is necessary to carry the same current as the rectangular bar?

Solution:

(1) $1.5'' = 1.5 \times 1000 = 1500$ mils

$0.25'' = 0.25 \times 1000 = 250$ mils

$A = 1500 \times 250 = 375{,}000$ sq mils

(2) To carry the same current, the cross-sectional area of the rectangular bar and the cross-sectional area of the round conductor must be equal. There are more circular mils than square mils in this area and, therefore,

$$A = \frac{375{,}000}{0.785} = 477{,}700 \text{ circular mils}$$

IMPORTANT **Note:** *Many electric cables are composed of stranded wires. The strands are usually single wires twisted together in sufficient numbers to make up the necessary cross-sectional area of the cable. The total area in circular mils is determined by multiplying the area of one strand in circular mils by the number of strands in the cable.*

4.2.3 CIRCULAR-MIL-FOOT

As shown in Figure 4.2, a *circular-mil-foot* is actually a unit of volume. More specifically, it is a unit conductor 1 foot in length and having a cross-sectional area of 1 circular mil. The circular-mil-foot is useful in making comparisons between wires that are made of different metals because it is considered a unit conductor. Because it is considered a unit

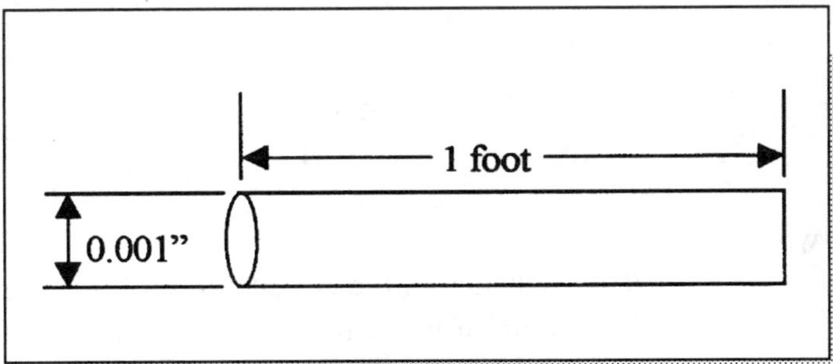

Figure 4.2
Circular-mil-foot.

conductor, the circular-mil-foot is useful in making comparisons between wires that are made of different metals. For example, a basis of comparison of the **resistivity** (see Section 4.3) of various substances may be made by determining the resistance of a circular-mil-foot of each of the substances.

IMPORTANT **Note:** *It is sometimes more convenient to employ a different unit of volume when working with certain substances. Accordingly, unit volume may also be taken as the centimeter cube. The inch cube may also be used. The unit of volume employed is given in tables of specific resistances.*

4.3 RESISTIVITY

All materials differ in their atomic structure and, therefore, in their ability to resist the flow of an electric current. The measure of the ability of a specific material to resist the flow of electricity is called its *resistivity,* or specific resistance—the resistance in ohms offered by unit volume (the circular-mil-foot) of a substance to the flow of electric current. Resistivity is the reciprocal of conductivity (i.e., the ease by which current flows in a conductor). A substance that has a high resistivity will have a low conductivity, and vice versa.

The resistance of a given length, for any conductor, depends upon the resistivity of the material, the length of the wire, and the cross-sectional area of the wire according to the equation

$$R = \rho \frac{L}{A} \tag{4.3}$$

where

R = resistance of the conductor, Ω
L = length of the wire, ft
A = cross-sectional area of the wire, CM
ρ = specific resistance or resistivity, CM x Ω/ft

TABLE 4.1. Resistivity (Specific Resistance).	
Substance	**Specific Resistance @ 20° (CM ft Ω)**
Silver	9.8
Copper (drawn)	10.37
Gold	4.7
Aluminum	17.02
Tungsten	33.2
Brass	42.1
Steel (soft)	95.8
Nichrome	660.0

The factor ρ (Greek letter rho, pronounced "roe") permits different materials to be compared for resistance according to their nature without regard to different lengths or areas. Higher values of ρ mean more resistance.

IMPORTANT

Key Point: *The resistivity of a substance is the resistance of a unit volume of that substance.*

Many tables of resistivity are based on the resistance in ohms of a volume of the substance 1 foot long and 1 circular mil in cross-sectional area. The temperature at which the resistance measurement is made is also specified. If the kind of metal of which the conductor is made is known, the resistivity of the metal may be obtained from a table. The resistivity, or specific resistance, of some common substances is given in Table 4.1.

IMPORTANT

Note: *Because silver, copper, gold, and aluminum have the lowest values of resistivity, they are the best conductors. Tungsten and iron have a much higher resistivity.*

EXAMPLE 4.4

Problem: What is the resistance of 1000 feet of copper wire having a cross-sectional area of 10,400 circular mils (No. 10 wire), the wire temperature being 20°C?

Solution: The resistivity (specific resistance), from Table 4.1, is 10.37. Substituting the known values in the preceding Equation (4.3), the resistance, *R*, is determined as

$$R = \rho \frac{L}{A} = 10.37 \times = \frac{1000}{10{,}400} \sim 1 \text{ ohm}$$

4.4 WIRE MEASUREMENT

Wires are manufactured in sizes numbered according to a table known as the American wire gage (AWG). Table 4.2 lists the standard wire sizes that correspond to the AWG. The gage numbers specify the size of round wire in terms of its diameter and cross-sectional area. Note the following:

- As the gage numbers increase from 1 to 40, the diameter and circular area decrease. Higher gage numbers mean smaller wire sizes. Thus, No. 12 is a smaller wire than No. 4.
- The circular area doubles for every three gage sizes. For example, No. 12 wire has about twice the area of No. 15 wire.
- The higher the gage number and the smaller the wire, the greater the resistance of the wire for any given length. Therefore, for 1000 ft of wire, No. 12 has a resistance of 1.62 Ω while No. 4 has a resistance of $0.253\ \Omega$.

4.5 FACTORS GOVERNING THE SELECTION OF WIRE SIZE

Several factors must be considered in selecting the size of wire to be used for transmitting and distributing electric power. These factors include allowable power loss in the line, the permissible voltage drop in

TABLE 4.2. Copper Wire Table.			
Gage #	Diameter	Circular mils	Ohms/1000 ft @ 25°C
1	289.0	83,700.0	.126
2	258.0	66,400.0	.159
3	229.0	52,600.0	.201
4	204.0	41,700.0	.253
5	182.0	33,100.0	.319
6	162.0	26,300.0	.403
7	144.0	20,800.0	.508
8	128.0	16,500.0	.641
9	114.0	13,100.0	.808
10	102.0	10,400.0	1.02
11	91.0	8230.0	1.28
12	81.0	6530.0	1.62
13	72.0	5180.0	2.04
14	64.0	4110.0	2.58
15	57.0	3260.0	3.25
16	51.0	2580.0	4.09
17	45.0	2050.0	5.16
18	40.0	1620.0	6.51
19	36.0	1290.0	8.21
20	32.0	1020.0	10.4
21	28.5	810.0	13.1
22	25.3	642.0	16.5
23	22.6	509.0	20.8
24	20.1	404.0	26.4
25	17.9	320.0	33.0
26	15.9	254.0	41.6
27	14.2	202.0	52.5
28	12.6	160.0	66.2
29	11.3	127.0	83.4
30	10.0	101.0	105.0
31	8.9	79.7	133.0
32	8.0	63.2	167.0
33	7.1	50.1	211.0
34	6.3	39.8	266.0
35	5.6	31.5	335.0
36	5.0	25.0	423.0
37	4.5	19.8	533.0
38	4.0	15.7	673.0
39	3.5	12.5	848.0
40	3.1	9.9	1070.0

the line, the current-carrying capacity of the line, and the ambient temperatures in which the wire is to be used.

1. *Allowable power loss* (I^2R) *in the line.* This loss represents electrical energy converted into heat. The use of large conductors will reduce the resistance and therefore the I^2R loss. However, large conductors are heavier and require more substantial supports; thus, they are more expensive initially than small ones.
2. *Permissible voltage drop (*IR *drop) in the line.* If the source maintains a constant voltage at the input to the line, any variation in the load on the line will cause a variation in line current and a consequent variation in the IR drop in the line. A wide variation in the IR drop in the line causes poor voltage regulation at the load.
3. *The current-carrying capacity of the line.* When current is drawn through the line, heat is generated. The temperature of the line will rise until the heat radiated, or otherwise dissipated, is equal to the heat generated by the passage of current through the line. If the conductor is insulated, the heat generated in the conductor is not so readily removed as it would be if the conductor were not insulated.
4. *Conductors installed where ambient temperature is relatively high.* When installed in such surroundings, the heat generated by external sources constitutes an appreciable part of the total conductor heating. Due allowance must be made for the influence of external heating on the allowable conductor current, and each case has its own specific limitations.

4.5.1 COPPER VS. OTHER METAL CONDUCTORS

If it were not cost prohibitive, silver, the best conductor of electron flow (electricity), would be the conductor of choice in electrical systems. Instead, silver is used only in special circuits where a substance with high conductivity is required.

The two most generally used conductors are copper and aluminum. Each has characteristics that make its use advantageous under certain circumstances. Likewise, each has certain disadvantages, or limitations.

Copper has a higher conductivity, and it is more ductile (can be drawn out into wire), has relatively high tensile strength, and can be easily soldered. It is more expensive and heavier than aluminum.

Aluminum has only about 60 percent of the conductivity of copper, but its lightness makes possible long spans, and its relatively large diameter for a given conductivity reduces corona (i.e., the discharge of electricity from the wire when it has a high potential). The discharge is greater when smaller diameter wire is used than when larger diameter wire is used. However, aluminum conductors are not easily soldered, and aluminum's relatively large size for a given conductance does not permit the economical use of an insulation covering.

IMPORTANT

Note: *Recent practice involves using copper wiring (instead of aluminum wiring) in house and some industrial applications. This is the case because aluminum connections are not as easily made as they are with copper. In addition, over the years, many fires have been started because of improperly connected aluminum wiring (i.e., poor connections = high resistance connections, resulting in excessive heat generation).*

A comparison of some of the characteristics of copper and aluminum is given in Table 4.3.

TABLE 4.3. Characteristics of Copper and Aluminum.

Characteristics	Copper	Aluminum
Tensile strength (lb/in²)	55,000	25,000
Tensile strength for same conductivity (lb)	55,000	40,000
Weight for same conductivity (lb)	100	48
Cross section for same conductivity (CM)	100	160
Specific resistance (Ω/mil ft)	10.6	17

4.6 TEMPERATURE COEFFICIENT

The resistance of pure metals—such as silver, copper, and aluminum—increases as the temperature increases. The *temperature coefficient* of resistance, α (Greek letter alpha), indicates how much the resistance changes for a change in temperature. A positive value for α means *R* increases with temperature; a negative α means *R* decreases; and a zero α means *R* is constant, not varying with changes in temperature. Typical values of α are listed in Table 4.4.

The amount of increase in the resistance of a 1-ohm sample of the copper conductor per degree rise in temperature (i.e., the temperature coefficient of resistance) is approximately 0.004. For pure metals, the temperature coefficient of resistance ranges between 0.004 and 0.006 ohm.

Thus, a copper wire having a resistance of 50 ohms at an initial temperature of 0°C will have an increase in resistance of 50 x 0.004, or 0.2 ohms (approximate) for the entire length of wire for each degree of temperature rise above 0°C. At 20°C, the increase in resistance is approximately 20 x 0.2, or 4 ohms. The total resistance at 20°C is 50 + 4, or 54 ohms.

TABLE 4.4 Properties of Conducting Materials (Approximate).

Material	Temperature Coefficient, Ω/°C, ?
Aluminum	0.004
Carbon	−0.0003
Constantan	0 (average)
Copper	0.004
Gold	0.004
Iron	0.006
Nichrome	0.0002
Nickel	0.005
Silver	0.004
Tungsten	0.005

IMPORTANT **Note:** *As shown in Table 4.4, carbon has a negative temperature coefficient. In general, α is negative for all semiconductors, such as germanium and silicon. A negative value for α means less resistance at higher temperatures. Therefore, the resistance of semiconductor diodes and transistors can be reduced considerably when they become hot with normal load current. Observe, also, that constantan has a value of zero for α (Table 4.4). Thus, it can be used for precision wire-wound resistors, which do not change resistance when the temperature increases.*

4.7 CONDUCTOR INSULATION

Electric current must be contained; it must be channeled from the power source to a useful load—safely. To accomplish this, electric current must be forced to flow only where it is needed. Moreover, current-carrying conductors must not be allowed (generally) to come in contact with one another, their supporting hardware, or personnel working near them. To accomplish this, conductors are coated or wrapped with various materials. These materials have such a high resistance that they are, for all practical purposes, nonconductors. They are generally referred to as *insulators* or *insulating materials.*

There is a wide variety of insulated conductors available to meet the requirements of any job. However, only the necessary minimum of insulation is applied for any particular type of cable designed to do a specific job. This is the case because insulation is expensive, has a stiffening effect, and is required to meet a great variety of physical and electrical conditions.

Two fundamental but distinctly different properties of insulation materials (e.g., rubber, glass, asbestos, and plastics) are insulation resistance and dielectric strength.

- *Insulation resistance* is the resistance to current leakage through and over the surface of insulation materials.
- *Dielectric strength* is the ability of the insulator to withstand potential difference and is usually expressed in terms of the voltage at which the insulation fails because of the electrostatic stress.

Various types of materials are used to provide insulation for electric conductors, including rubber, plastics, varnished cloth, paper, silk, cotton, and enamel.

4.8 CONDUCTOR SPLICES AND TERMINAL CONNECTIONS

When conductors join each other, or connect to a load, *splices* or *terminals* must be used. It is important that they be properly made, because any electric circuit is only as good as its weakest connection. The basic requirement of any splice or connection is that it be both mechanically and electrically as strong as the conductor or device with which it is used. High quality workmanship and materials must be employed to ensure lasting electrical contact, physical strength, and insulation (if required).

IMPORTANT

Important Point: *Conductor splices and connections are an essential part of any electric circuit.*

4.9 SOLDERING OPERATIONS

Soldering operations are a vital part of electrical and/or electronics maintenance procedures. Soldering is a manual skill that must be learned by all personnel who work in the field of electricity. Obviously, practice is required to develop proficiency in the techniques of soldering.

In performing a soldering operation, both the solder and the material to be soldered (e.g., electric wire and/or terminal lugs) must be heated to a temperature that allows the solder to flow. If either is heated inadequately, **cold** solder joints result (i.e., high resistance connections are created). Such joints do not provide either the physical strength or the electrical conductivity required. Moreover, in soldering operations, it is necessary to select a solder that will flow at a temperature low enough to avoid damage to the part being soldered or to any other part or material in the immediate vicinity.

4.10 SOLDERLESS CONNECTORS

Generally, terminal lugs and splicers that do not require solder are more widely used (because they are easier to mount correctly) than those that do require solder. Solderless connectors—made in a wide variety of sizes and shapes—are attached to their conductors by means of several different devices, but the principle of each is essentially the same. They are all crimped (squeezed) tightly onto their conductors. They afford adequate electrical contact, plus great mechanical strength.

4.11 INSULATION TAPE

The carpenter has a saw, the dentist has pliers, the plumber has a wrench, and the electrician has insulating tape. Accordingly, one of the first things the rookie maintenance operator learns (a rookie who is also learning proper and safe techniques for performing electrical work) is the value of electrical insulation tape. Normally, the use of electrical insulating tape comes into play as the final step in completing a splice or joint, to place insulation over the bare wire at the connection point.

Typically, insulation tape used should be the same basic substance as the original insulation, usually a rubber-splicing compound. When using rubber (latex) tape as the splicing compound where the original insulation was rubber, it should be applied to the splice with a light tension so that each layer presses tightly against the one underneath it. In addition to the rubber tape application (which restores the insulation to original form), restoring with friction tape is also often necessary.

In recent years, plastic electrical tape has come into wide use. It has certain advantages over rubber and friction tape. For example, it will withstand higher voltages for a given thickness. Single thin layers of certain commercially available plastic tape will stand several thousand volts without breaking down.

IMPORTANT

Important Point: *Be advised that though the use of plastic electrical tape has become almost universal in industrial applications, it must be applied in more layers—because it is thinner than rubber or friction tape—to ensure an extra margin of safety.*

Self-Test

4.1 What is the function of the wire conductor?

4.2 The resistance in ohms of a unit conductor or a given substance is called the ________________.

4.3 The ____________ ________ is the standard unit of wire cross-sectional area used in most wire tables.

4.4 Which is smaller, circular mil or square mil?

4.5 Resistivity is the reciprocal of __________________.

4.6 Number 8 wire is ___________ than number 4 wire.

4.7 For every three gage sizes, the circular area of a wire ______________.

4.8 The higher the gage number and the smaller the wire, the ____________ the resistance of the wire for _____ given length.

4.9 List the four factors that govern the selection of wire size.

4.10 Carbon has a ____________ temperature coefficient.

4.11 Define dielectric strength:

4.12 A cold solder joint is the same as a ________ resistance connection.

4.13 Typically, insulation tape used should be the same basic substance as the ____________ insulation.

4.14 The biggest advantage of using plastic insulation tape over other types of tape is____________.

4.15 What is the resistance of 500 ft of No. 20 copper wire?

Electromagnetism

TOPICS

Magnetic Field around a Single Conductor
Magnetic Field of a Coil
Magnetic Units
Properties of Magnetic Materials
Electromagnets

5.1 INTRODUCTION

In Chapter 1 (Section 1.6), we discussed the fundamental theories concerning simple magnets and magnetism. Those discussions dealt mainly with forms of magnetism that were not related directly to electricity—permanent magnets for instance. Further, only brief mention was made of those forms of magnetism having direct relation to electricity—producing electricity with magnetism for instance.

IMPORTANT **Note:** *In this chapter, we resume our discussion of magnetism and build upon the information presented in Chapter 1. Because we begin this chapter with a more advanced discussion of magnetism, it may be beneficial for you to review parts of Chapter 1 from time to time.*

In medicine, anatomy and physiology are so closely related that the medical student cannot study one at length without involving the other. A similar relationship holds for the electrical field; that is, magnetism and basic electricity are so closely related that one cannot be studied at length without involving the other. This close fundamental relationship is continually borne out in subsequent chapters of this manual, such as in the study of generators, transformers, and motors. To be proficient in electricity, the maintenance operator must become familiar with such general relationships that exist between magnetism and electricity as follows:

- Electric current flow will always produce some form of magnetism.

Key Terms Used in This Chapter

MAGNETIC FLUX	Similar to current flow in Ohm's Law and comprises the total number of lines of force existing in the magnetic circuit.
MAGNETOMOTIVE FORCE	Force that produces flux in a magnetic circuit.
HYSTERESIS	Condition whereby magnetic flux in an iron core lags behind the increase or decrease of the magnetizing force.

- Magnetism is by far the most commonly used means for producing or using electricity.
- The peculiar behavior of electricity under certain conditions is caused by magnetic influences.

5.2 MAGNETIC FIELD AROUND A SINGLE CONDUCTOR

In 1819, Hans Christian Oersted, a Danish scientist, discovered that a field of magnetic force exists around a single wire conductor carrying an electric current. In Figure 5.1, a wire is passed through a piece of cardboard and is connected through a switch to a dry cell. With the switch open (no current flowing), if we sprinkle iron filings on the cardboard, then tap it gently, the filings will fall back haphazardly. Now, if we close the switch, current will begin to flow in the wire. If we tap the cardboard again, the magnetic effect of the current in the wire will cause the filings to fall back into a definite pattern of concentric circles with the wire as the center of the circles. Every section of the wire has this field of force around it in a plane perpendicular to the wire, as shown in Figure 5.2.

The ability of the magnetic field to attract bits of iron (as demonstrated in Figure 5.1) depends on the number of lines of force present. The

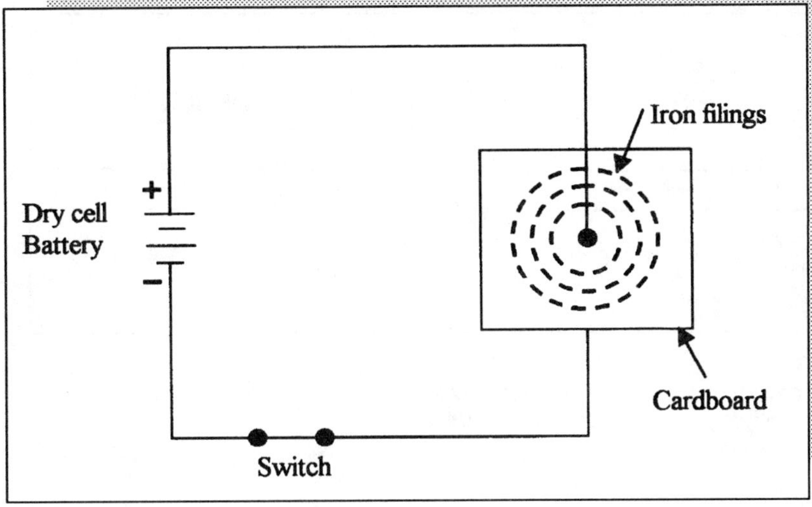

Figure 5.1
A circular pattern of magnetic force exists around a wire carrying an electric current.

strength of the magnetic field around a wire carrying a current depends on the current, because it is the current that produces the field. The greater the current, the greater the strength of the field. A large current will produce many lines of force extending far from the wire, while a small current will produce only a few lines close to the wire, as shown in Figure 5.3.

Figure 5.2
The circular fields of force around a wire carrying a current are in planes that are perpendicular to the wire.

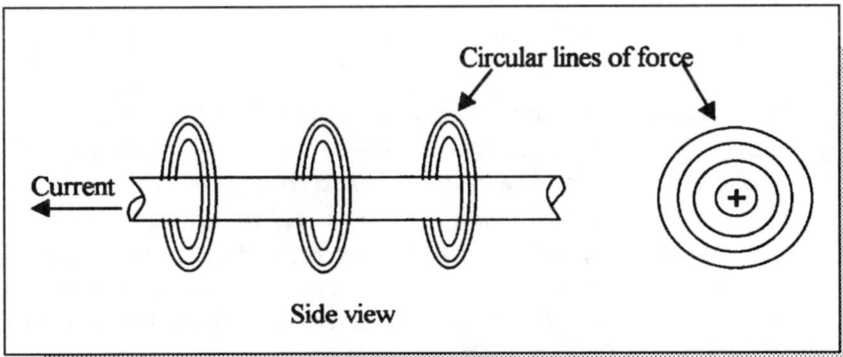

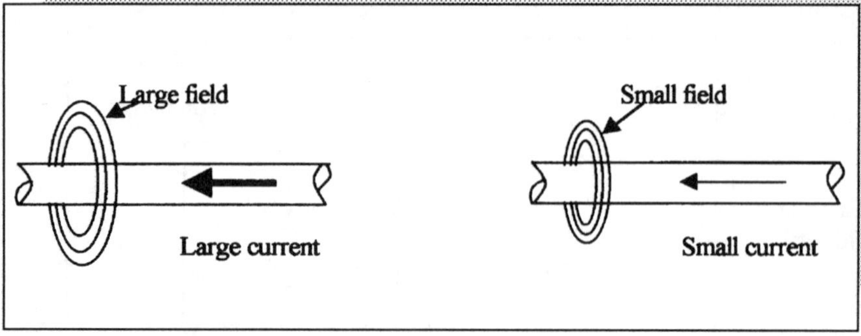

Figure 5.3
The strength of the magnetic field around a wire carrying a current depends on the amount of current.

5.2.1 POLARITY OF A SINGLE CONDUCTOR

The relation between the direction of the magnetic lines of force around a conductor and the direction of current flow along the conductor may be determined by means of the **left-hand rule for a conductor.** If the conductor is grasped in the left hand with the thumb extended in the direction of electron flow (– to +), the fingers will point in the direction of the magnetic lines of force. This is the same direction that the north pole of a compass would point if the compass were placed in the magnetic field.

IMPORTANT

Important Note: *Arrows are generally used in electric diagrams to denote the direction of current flow along the length of wire. Where cross sections of wire are shown, a special view of the arrow is used. A cross-sectional view of a conductor that is carrying current toward the observer is illustrated in Figure 5.4(a). The direction of current is indicated by a dot, which represents the head of the arrow. A conductor that is carrying current away from the observer is illustrated in Figure 5.4(b). The direction of current is indicated by a cross, which represents the tail of the arrow.*

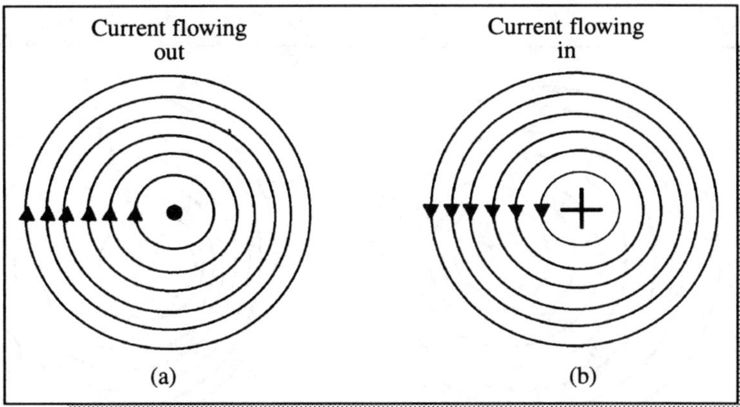

Figure 5.4
Magnetic field around a current-carrying conductor.

5.2.2 FIELD AROUND TWO PARALLEL CONDUCTORS

When two parallel conductors carry current in the same direction, the magnetic fields tend to encircle both conductors, drawing them together with a force of attraction, as shown in Figure 5.5(a). Two parallel conductors carrying currents in opposite directions are shown in Figure 5.5(b). The field around one conductor is opposite in direction to the field around the other conductor. The resulting lines of force are crowded

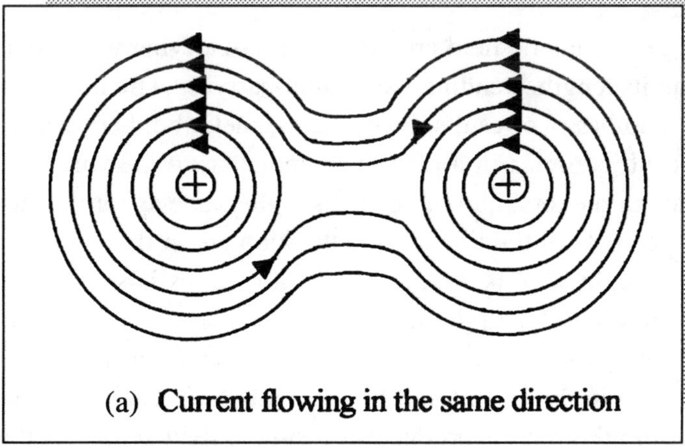

Figure 5.5
Magnetic field around two parallel conductors.

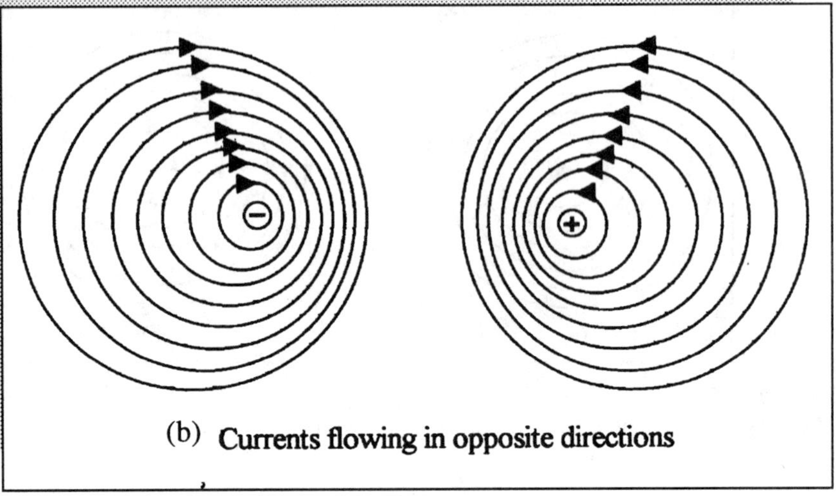

Figure 5.5 (continued)
Magnetic field around two parallel conductors.

together in the space between the wires and tend to push the wires apart. Therefore, two parallel adjacent conductors carrying currents in the same direction attract each other, and two parallel conductors carrying currents in opposite directions repel each other.

5.3 MAGNETIC FIELD OF A COIL

The magnetic field around a current-carrying wire exists at all points along its length. Bending the current-carrying wire into the form of a single loop has two results. First, the magnetic field consists of more dense concentric circles in a plane perpendicular to the wire (see Figure 5.2), although the total number of lines is the same as for the straight conductor. Second, all the lines inside the loop are in the same direction. When this straight wire is wound around a core, as is shown in Figure 5.6, it becomes a coil, and the magnetic field assumes a different shape. When current is passed through the coiled conductor, the magnetic field of each turn of wire links with the fields of adjacent turns. The combined influence of all the turns produces a two-pole field similar to that of a simple bar magnet. One end of the coil will be a north pole, and the other end will be a south pole.

5.3.1 POLARITY OF AN ELECTROMAGNETIC COIL

In Figure 5.3, it was shown that the direction of the magnetic field around straight conductor depends on the direction of current flow through that conductor. Thus, a reversal of current flow through a conductor causes a reversal in the direction of the magnetic field that is produced. It follows that a reversal of the current flow through a coil also causes a reversal of its two-pole field. This is true because that field is the product of the linkage between the individual turns of wire on the coil. Therefore, if the field of each turn is reversed, it follows that the total field (coils' field) is also reversed.

When the direction of electron flow through a coil is known, its polarity may be determined by use of the **left-hand rule for coils.** This rule is illustrated in Figure 5.6 and is stated as follows: Grasping the coil in the left hand, with the fingers "wrapped around" in the direction of electron flow, the thumb will point toward the north pole.

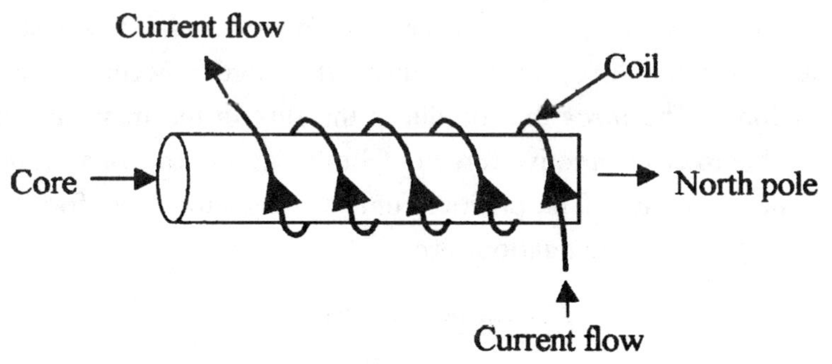

Figure 5.6
Current-carrying coil.

5.3.2 STRENGTH OF AN ELECTROMAGNETIC FIELD

The strength, or intensity, of the magnetic field of a coil depends on a number of factors.

- the *number of turns* of conductor.
- the *amount of current flow* through the coil.
- the *ratio of the coil's length to its width.*
- the *type of material in the core.*

5.4 MAGNETIC UNITS

The law of current flow in the electric circuit is similar to the law for the establishing of flux in the magnetic circuit.

The magnetic flux, ϕ (phi), is similar to current in the Ohm's Law formula and comprises the total number of lines of force existing in the magnetic circuit. The **maxwell** is the unit of flux—that is, 1 line of force is equal to 1 maxwell.

IMPORTANT

Note: *The maxwell is often referred to as simply a line of force, line of induction, or line.*

The *strength* of a magnetic field in a coil of wire depends on how much current flows in the turns of the coil. The more current, the stronger the magnetic field. Also, the more turns, the more concentrated are the lines of force. The *force* that produces the flux in the magnetic circuit (comparable to electromotive force in Ohm's Law) is known as *magnetomotive force*, or mmf. The practical unit of magnetomotive force is the **ampere-turn** (At). In equation form,

$$\boldsymbol{F = \text{ampere-turns} = NI} \tag{5.1}$$

where

F = magnetomotive force, At
N = number of turns
I = current, A

EXAMPLE 5.1

Problem: Calculate the ampere-turns for a coil with 2000 turns and a 5-Ma current.

Solution: Use Equation (5.1), and substitute

$$N = 2000 \text{ and } I = 5 \times 10^{-3} \text{ A}$$
$$NI = 2000(5 \times 10^{-3}) = 10 \text{ At}$$

The unit of *intensity* of magnetizing force per unit of length is designated as *H* and is sometimes expressed as Gilberts per centimeter of length. Expressed as an equation,

$$H = \frac{NI}{L} \qquad (5.2)$$

where

H = magnetic field intensity, ampere-turns per meter (At/m)

NI = ampere-turns, At

L = length between poles of the coil, m

IMPORTANT **Note:** *Equation (5.2) is for a solenoid.* H *is the intensity of an air core. With an iron core,* H *is the intensity through the entire core, and* L *is the length or distance between poles of the iron core.*

5.5 PROPERTIES OF MAGNETIC MATERIALS

In this section, we discuss two important properties of magnetic materials: permeability and hysteresis.

5.5.1 PERMEABILITY

When the core of an electromagnet is made of annealed sheet steel, it produces a stronger magnet than if a cast iron core is used. This is the case because annealed sheet steel is more readily acted upon by the magnetizing force of the coil than is the hard cast iron. Simply put, soft sheet steel is said to have greater *permeability* because of the greater ease with which magnetic lines are established in it.

Recall from Section (1.6.1) that permeability is the relative ease with which a substance conducts magnetic lines of force. The permeability of air is arbitrarily set at 1. The permeability of other substances is the ratio of their ability to conduct magnetic lines compared to that of air. The permeability of nonmagnetic materials, such as aluminum, copper, wood, and brass, is essentially unity, or the same as for air.

IMPORTANT ***Important Note:*** *The permeability of magnetic materials varies with the degree of magnetization, being smaller for high values of flux density.*

IMPORTANT ***Important Note:*** Reluctance, *which is analogous to resistance, is the opposition to the production of flux in a material and is inversely proportional to permeability. Iron has high permeability and, therefore, low reluctance. Air has low permeability and, hence, high reluctance.*

5.5.2 HYSTERESIS

When the current in a coil of wire reverses thousands of times per second, a considerable loss of energy can occur. This loss of energy is caused by *hysteresis*. Hysteresis means "a lagging behind"; that is, the magnetic flux in an iron core lags behind the increases or decreases of the magnetizing force.

The simplest method of illustrating the property of hysteresis is by graphical means, such as the hysteresis loop shown in Figure 5.7.

The hysteresis loop (Figure 5.7) is a series of curves that show the characteristics of a magnetic material. Opposite directions of current result are in the opposite directions of $+H$ and $-H$ for field intensity. Similarly, opposite polarities are shown for flux density as $+B$ and $-B$. The current starts at the center 0 (zero) when the material is unmagnetized. Positive H values increase B to saturation at $+B_{max}$. Next, H decreases

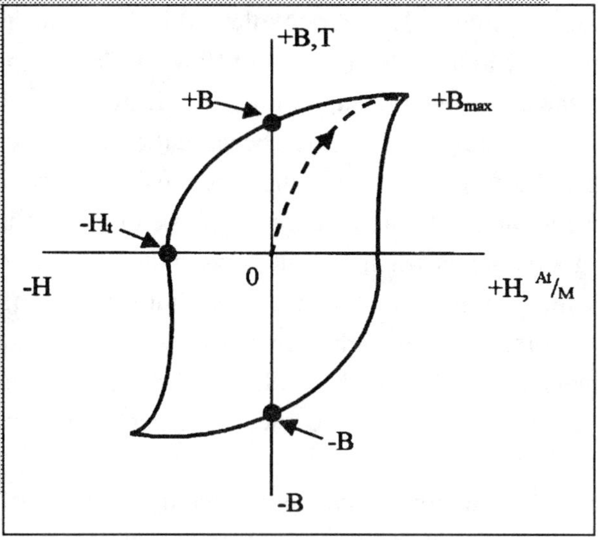

Figure 5.7
Hysteresis loop.

to zero, but B drops to the value of B_r because of hysteresis. The current that produced the original magnetization now is reversed so that H becomes negative. B drops to zero and continues to $-B_{max}$. As the $-H$ values decrease, B is reduced to $-B_r$ when H is zero. Now, with a positive swing of current, H becomes positive, producing saturation at $+B_{max}$ again. The hysteresis loop is now completed. The curve doesn't return to zero at the center because of hysteresis.

5.6 ELECTROMAGNETS

An *electromagnet* is composed of a coil of wire wound around a core that is normally soft iron, because of its high permeability and low hysteresis. When direct current flows through the coil, the core will become magnetized with the same polarity that the coil would have without the core. If the current is reversed, the polarity of the coil and core are reversed.

The electromagnet is of great importance in electricity simply because the magnetism can be turned on or turned off at will. The starter solenoid (an electromagnet) in automobiles and power boats is a good example. In an automobile or boat, an electromagnet is part of a relay that connects the battery to the induction coil, which generates the very high voltage needed to start the engine. The starter solenoid isolates this high voltage from the ignition switch. When no current flows in the coil, it is an air core, but when the coil is energized, a movable soft-iron core does two things. First, the magnetic flux is increased because the soft-iron core is more permeable than the air core. Second, the flux is more highly concentrated. All this concentration of magnetic lines of force in the soft-iron core results in a very good magnet when current flows in the coil, but soft-iron loses its magnetism quickly when the current is shut off. The effect of the soft iron is, of course, the same whether it is movable, as in some solenoids, or permanently installed in the coil. An electromagnet then consists basically of a coil and a core; it becomes a magnet when current flows through the coil.

The ability to control the action of magnetic force makes an electromagnet very useful in many circuit applications. Many of the applications of electromagnets are discussed throughout this manual.

Self-Test

5.1 What is the relationship between flux density and magnetic field strength?

5.2 How can you determine the direction of the magnetic field around a conductor when the direction of current flow is known?

5.3 Permeability depends on what two factors?

5.4 What is hysteresis?

5.5 Name three ways the magnetic field strength of a current-carrying coil can be increased.

5.6 When does an electromagnet actually function as a magnet?

5.7 Reluctance is inversely proportional to ________________.

5.8 At/m = ________________.

5.9 *NI* = ______________.

5.10 Lines of force flow from ________________.

A-C Theory

TOPICS

Basic A-C Generator
Characteristic Values of A-C Voltages and Current
Resistance in A-C Circuits

6.1 INTRODUCTION

Because voltage is induced in a conductor when lines of force are cut, the amount of the induced emf depends on the number of lines cut in a unit of time. To induce an emf of 1 volt, a conductor must cut 100 million lines of force per second. To obtain this great number of "cuttings," the conductor is formed into a loop and is rotated on an axis at great speed (see Figure 6.1). The two sides of the loop become individual conductors in series, each side of the loop cutting lines of force and inducing twice the voltage that a single conductor would induce. In commercial generators, the number of "cuttings" and the resulting emf are increased by (1) increasing the number of lines of force by using more magnets or stronger electromagnets, (2) using more conductors or loops, and (3) rotating the loops faster. (Note: both a-c and d-c generators are covered in Chapter 11.)

How an a-c generator operates to produce an a-c voltage and current is a basic concept today and is taught in elementary and middle school science classes. Of course, we accept technological advances as commonplace, today—we surf the Internet, watch cable television, use our cell phones, and take space flight as a given—and consider producing the electricity that makes all these technologies possible as our right. These technologies are bottom shelf to us today—we have them available to us so we simply use them. This point of view surely was not held initially—especially by those who broke ground in developing technology and electricity.

In groundbreaking years of electric technology development, the geniuses of the science of electricity (including George Simon Ohm) performed their technological breakthroughs in faltering steps. We tend to

Key Terms Used in This Chapter

ALTERNATING CURRENT (A-C)	An electric current that reverses direction periodically, usually many times per second.
ALTERNATION	A variation, either positive or negative, of a wave-form from zero to maximum and back to zero, equaling one-half a cycle.
HERTZ	A unit of frequency; a periodic oscillation has a frequency of *n* hertz if in 1 second it goes through *n* cycles.
FREQUENCY	The number of cycles completed by a periodic quantity in a unit time (e.g., the number of complete cycles that occur in one second).
WAVELENGTH	The distance traveled by an a-c wave during one complete cycle.
PHASE ANGLE	Refers to the value of the electrical angle at any point during the cycle.
VECTOR QUANTITY	Is one that has magnitude and direction.

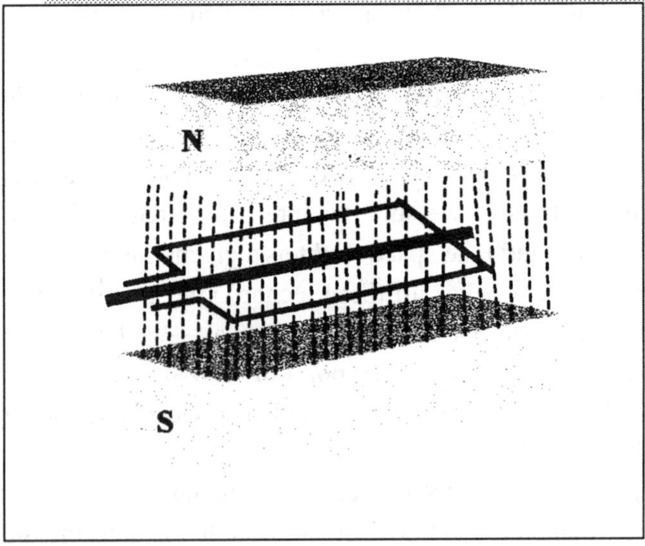

Figure 6.1
Loop rotating in magnetic field produces an a-c voltage.

forget that those first faltering steps of scientific achievement in the field of electricity were performed with crude and, for the most part, homemade apparatus.

Indeed, the innovators of electricity had to fabricate nearly all the laboratory equipment used in their experiments. At the time, the only convenient source of electrical energy available to these early scientists was the voltaic cell, invented some years earlier. Because of the fact that cells and batteries were the only sources of power available, some of the early electrical devices were designed to operate from **direct current.**

Thus, initially, direct current was used extensively. However, when the use of electricity became widespread, certain disadvantages in the use of direct current became apparent. In a direct-current system, the supply voltage must be generated at the level required by the load. To operate a 240-volt lamp, for example, the generator must deliver 240 volts. A 120-volt lamp could not be operated from this generator by any convenient means. A resistor could be placed in series with the 120-volt lamp to drop the extra 120 volts, but the resistor would waste an amount of power equal to that consumed by the lamp.

Another disadvantage of direct-current systems is the large amount of power lost due to the resistance of the transmission wires used to carry current from the generating station to the consumer. This loss could be greatly reduced by operating the transmission line at very high voltage and low current. This is not a practical solution in a d-c system, however, because the load would also have to operate at high voltage. As a result of the difficulties encountered with direct current, practically all modern power distribution systems use **alternating current (a-c),** including water/wastewater treatment plants.

Unlike d-c voltage, a-c voltage can be stepped up or down by a device called a **transformer** (see Chapter 11). Transformers permit the transmission lines to be operated at high voltage and low current for maximum efficiency. Then, at the consumer end, the voltage is stepped down to whatever value the load requires by using a transformer. Due to its inherent advantages and versatility, alternating current has replaced direct current in all but a few commercial power distribution systems.

The purpose of this chapter is to provide the water/wastewater maintenance operator with the basic information required to understand basic a-c theory. Accordingly, we explain how an alternator generates an a-c voltage; explain frequency and state the factors that affect it; analyze the sine waveform of a-c voltage or current; make the mathematical conversions between maximum and effective values of voltage or current; and explain vectors, or phasors.

6.2 BASIC A-C GENERATOR

As shown in Figure 6.1, an a-c voltage and current can be produced when a conductor loop rotates through a magnetic field and cuts lines of force to generate an induced a-c voltage across its terminals. This describes the basic principle of operation of an alternating current generator, or **alternator.** An alternator converts mechanical energy into electrical energy. It does this by utilizing the principle of **electromagnetic induction.** The basic components of an alternator are an armature, about which many turns of conductor are wound, which rotates in a magnetic field, and some means of delivering the resulting alternating current to an external circuit. (Note: in Chapter 11, we go into generator construction in more detail; in this chapter, we concentrate on theory of operation).

6.2.1 CYCLE

An a-c voltage is one that continually changes in magnitude and periodically reverses in polarity (see Figure 6.2). The zero axis is a horizontal line across the center. The vertical variations on the voltage wave show

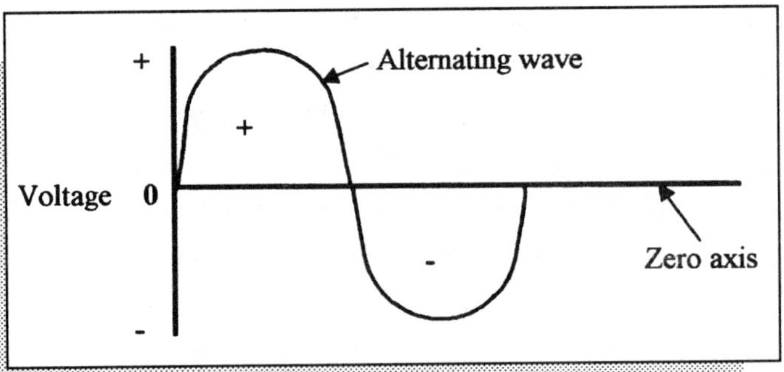

Figure 6.2
An a-c voltage waveform.

the changes in magnitude. The voltages above the horizontal axis have positive (+) polarity, while voltages below the horizontal axis have negative (–) polarity.

IMPORTANT **Note:** *To bring the important points presented up to this point into finer focus and to expand the presentation, Figure 6.3 is provided with accompanying explanation.*

Figure 6.3 shows a suspended loop of wire (conductor or armature) being rotated (moved) in a counterclockwise direction through the magnetic field between the poles of a permanent magnet. For ease of explanation, the loop has been divided into a thick and thin half. Notice that in part (a), the thick half is moving along (parallel to) the lines of force. Consequently, it is cutting none of these lines.The same is true of the thin half, moving in the opposite direction. Because the conductors are not cutting any lines of force, no emf is induced. As the loop rotates toward the position shown in part (b), it cuts more and more lines of force per second because it is cutting more directly across the field (lines of force) as it approaches the position shown in (b). At position (b) the induced voltage is greatest because the conductor is cutting directly across the field.

As the loop continues to be rotated toward the position shown in part (c), it cuts fewer and fewer lines of force per second. The induced voltage decreases from its peak value. Eventually, the loop is once again moving in a plane parallel to the magnetic field, and no voltage (zero voltage) is induced. The loop has now been rotated through half a circle (one alternation, or 180°). The sine curve shown in the lower part of Figure 6.3 shows the induced voltage at every instant of rotation of the loop. Notice that this curve contains 360°, or two alternations. **Two alternations represent one complete circle of rotation.**

IMPORTANT **Important Point:** *Two complete alternations in a period of time is called a* ***cycle.***

In Figure 6.3, if the loop is rotated at a steady rate, and if the strength of the magnetic field is uniform, the number of cycles per second (cps), or **hertz,** and the voltage will remain at fixed values. Continuous rotation will produce a series of sine-wave voltage cycles or, in other words, an a-c voltage. In this way, mechanical energy is converted into electrical energy.

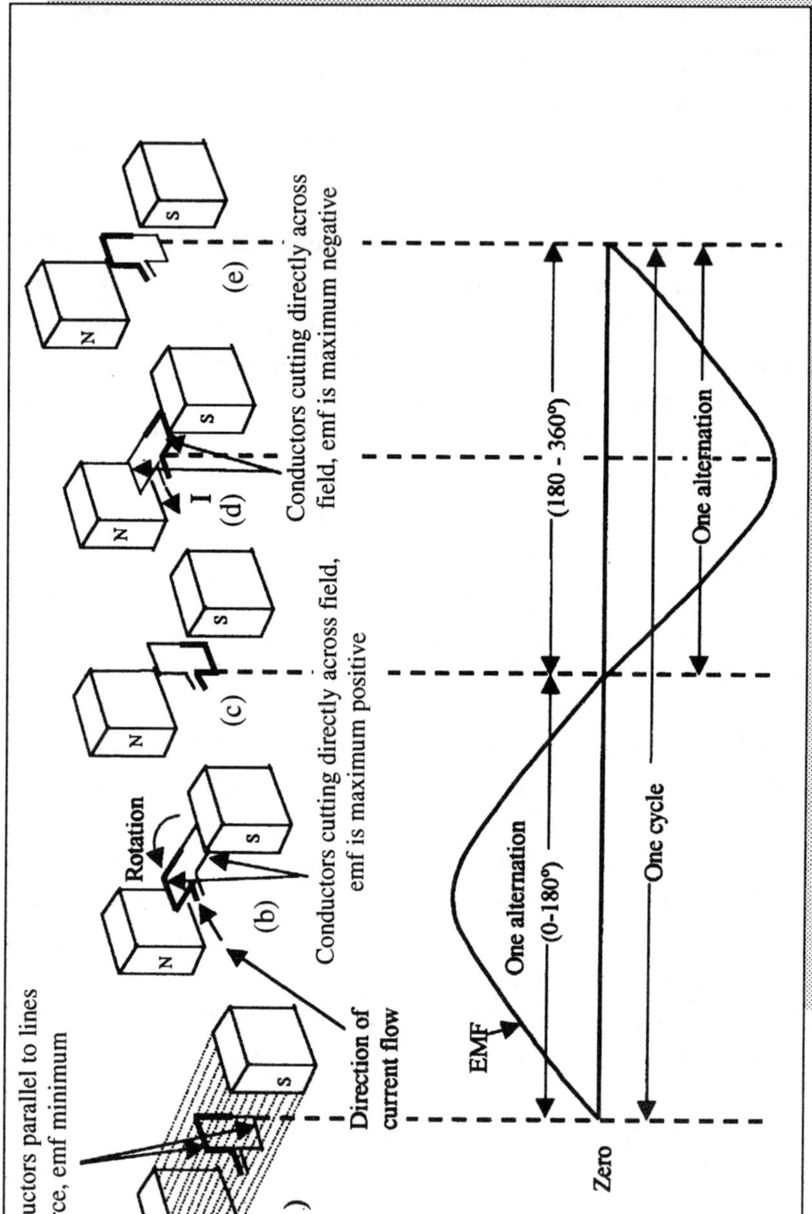

Figure 6.3
Basic alternating-current generator.

6.2.2 FREQUENCY, PERIOD, AND WAVELENGTH

The *frequency* of an alternating voltage or current is the number of complete cycles occurring in each second of time. It is indicated by the symbol f and is expressed in hertz (Hz). One cycle per second equals 1 hertz. Thus, 60 cycles per second (cps) equals 60 Hz. A frequency of 2 Hz [(Figure 6.4(a)] is twice the frequency of 1 Hz [Figure 6.4(b)].

The amount of time for the completion of one cycle is the *period.* It is indicated by the symbol T for time and is expressed in seconds (s). Frequency and period are reciprocals of each other.

$$f = \frac{1}{T} \tag{6.1}$$

$$T = \frac{1}{f} \tag{6.2}$$

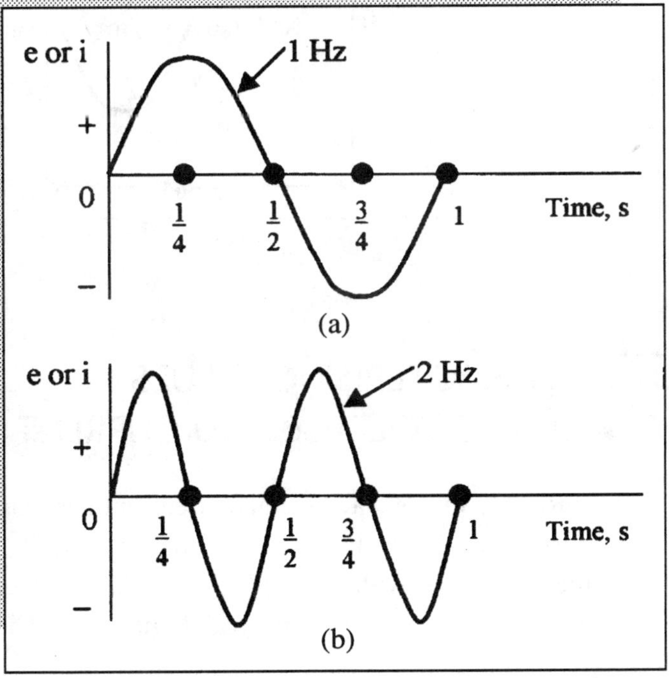

Figure 6.4 Comparison of frequencies.

Important Point: *The higher the frequency, the shorter the period.*

IMPORTANT

The angle of 360° represents the time for one cycle, or the period T. So, we can show the horizontal axis of the sine wave in units of either electrical degrees or seconds (see Figure 6.5).

The *wavelength* is the length of one complete wave or cycle. It depends upon the frequency of the periodic variation and its velocity of transmission. It is indicated by the symbol λ (Greek lowercase lambda). Expressed as a formula:

$$\lambda = \frac{\text{velocity}}{\text{frequency}} \tag{6.3}$$

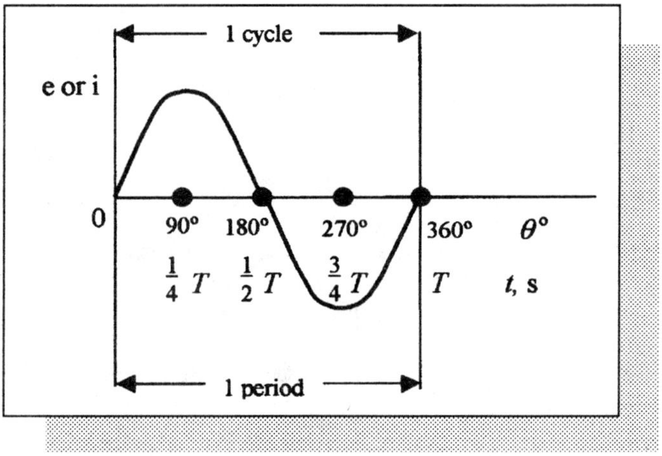

Figure 6.5 *Relationship between electrical degrees and time.*

6.3 CHARACTERISTIC VALUES OF A-C VOLTAGE AND CURRENT

Because an a-c sine wave voltage or current has many instantaneous values throughout the cycle, it is convenient to specify magnitudes for comparing one wave with another. The peak, average, or root-mean-square (rms) value can be specified (see Figure 6.6). These values apply to current or voltage.

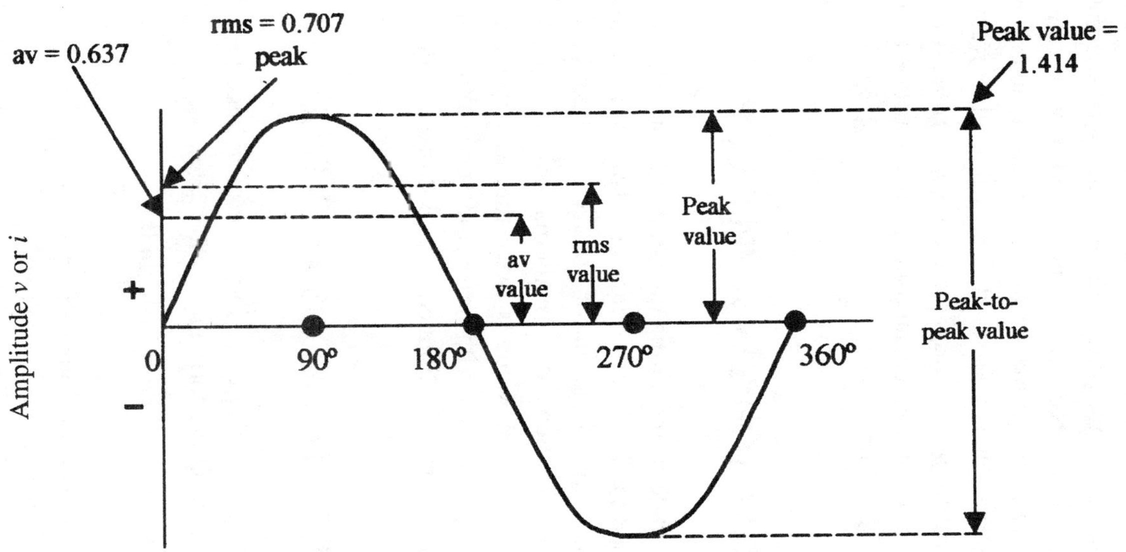

Figure 6.6
Amplitude values for a-c sine wave.

6.3.1 PEAK AMPLITUDE

One of the most frequently measured characteristics of a sine wave is its amplitude. Unlike d-c measurement, the amount of alternating current or voltage present in a circuit can be measured in various ways. In one method of measurement, the maximum amplitude of either the positive or the negative alternation is measured. The value of current or voltage obtained is called the *peak voltage* or the *peak current.* To measure the peak value of current or voltage, an oscilloscope must be used. The peak value is illustrated in Figure 6.6.

6.3.2 PEAK-TO-PEAK AMPLITUDE

A second method of indicating the amplitude of a sine wave consists of determining the total voltage or current between the positive and negative peaks. This value of current or voltage is called the *peak-to-peak value* (see Figure 6.6). Because both alternations of a pure sine wave are identical, the peak-to-peak value is twice the peak value. Peak-to-peak voltage is usually measured with an oscilloscope, although some voltmeters have a special scale calibrated in peak-to-peak volts.

6.3.3 INSTANTANEOUS AMPLITUDE

The *instantaneous value* of a sine wave of voltage for any angle of rotation is expressed by the formula:

$$e = E_m \times \sin \theta \qquad (6.4)$$

where

e = the instantaneous voltage
E_m = the maximum or peak voltage
$\sin \theta$ = the sine of angle at which e is desired

Similarly, the equation for the instantaneous value of a sine wave of current would be

$$i = I_m \times \sin \theta \qquad (6.5)$$

where

i = the instantaneous current
I_m = the maximum or peak current
$\sin \theta$ = the sine of the angle at which i is desired

IMPORTANT **Note:** *The instantaneous value of voltage constantly changes as the armature of an alternator moves through a complete rotation. Because current varies directly with voltage, according to Ohm's Law, the instantaneous changes in current also result in a sine wave whose positive and negative peaks and intermediate values can be plotted exactly as we plotted the voltage sine wave. However, instantaneous values are not useful in solving most a-c problems, so an* ***effective*** *value is used.*

6.3.4 EFFECTIVE OR RMS VALUE

The *effective value* of an a-c voltage or current of sine waveform is defined in terms of an equivalent heating effect of a direct current. Heating effect is independent of the direction of current flow.

IMPORTANT **Important Point:** *Because all instantaneous values of induced voltage are somewhere between zero and E_M (maximum, or peak voltage), the effective value of a sine wave voltage or current must be greater than zero and less than E_M (the maximum, or peak voltage).*

The alternating current of sine waveform having a maximum value of 14.14 amps produces the same amount of heat in a circuit having a resistance of 1 ohm as a direct current of 10 amps. Because this is true, we can work out a constant value for converting any peak value to a corresponding effective value. This constant is represented by X in the simple equation below. Solve for X to three decimal places.

$$14.14X = 10$$
$$X = 0.707$$

The effective value is also called the *root-mean-square (rms)* value because it is the square root of the average of the squared values between zero and maximum. The effective value of an a-c current is stated in terms of an equivalent d-c current. The phenomenon used as the standard comparison is the heating effect of the current.

IMPORTANT

Important Point: *Anytime an a-c voltage or current is stated without any qualifications, it is assumed to be an effective value.*

In many instances, it is necessary to convert from effective to peak or vice versa using a standard equation. Figure 6.6 shows that the peak value of a sine wave is 1.414 times the effective value; therefore, the equation we use is

$$E_m = E \times 1.414 \qquad (6.6)$$

where

E_m = maximum or peak voltage
E = effective or rms voltage

and

$$I_m = I \times 1.414 \qquad (6.7)$$

where

I_m = maximum or peak current
I = effective or rms current

Upon occasion, it is necessary to convert a peak value of current or voltage to an effective value. This is accomplished by using the following equations:

$$E = E_m \times 0.707 \qquad (6.8)$$

where

E = effective voltage
E_m = the maximum or peak voltage

$$I = I_m \times 0.707 \qquad (6.9)$$

where

I = the effective current
I_m = the maximum or peak current

6.3.5 AVERAGE VALUE

Because the positive alternation is identical to the negative alternation, the *average value* of a complete cycle of a sine wave is zero. In certain types of circuits, however, it is necessary to compute the average value of one alternation. Figure 6.6 shows that the average value of a sine wave is 0.637 x peak value and therefore:

$$\textbf{Average value} = \mathbf{0.637 \times Peak\ value} \qquad (6.10)$$

or

$$E_{avg} = E_m \times 0.637$$

where

E_{avg} = the average voltage of one alternation
E_m = the maximum or peak voltage

similarly

$$I_{avg} = I_m \times 0.637 \qquad (6.11)$$

where

I_{avg} = the average current in one alternation
I_m = the maximum or peak current

Table 6.1 lists the various values of sine wave amplitude used to multiply in the conversion of a-c sine wave voltage and current.

TABLE 6.1. A-C Sine Wave Conversion Table.

Multiply the Value	by	To Get the Value
Peak	2	Peak-to-peak
Peak-to-peak	0.5	Peak
Peak	0.637	Average
Average	1.637	Peak
Peak	0.707	rms (effective)
rms (effective)	1.414	Peak
Average	1.110	rms (effective)
rms (effective)	0.901	Average

6.4 RESISTANCE IN A-C CIRCUITS

If a sine wave of voltage is applied to a resistance, the resulting current will also be a sine wave. This follows Ohm's Law that states that the current is directly proportional to the applied voltage. Figure 6.7 shows a

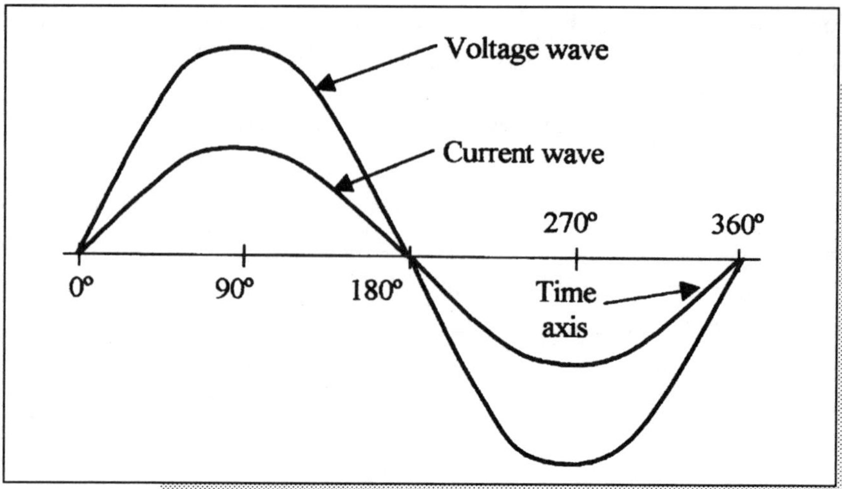

Figure 6.7
Voltage and current waves in phase.

sine wave of voltage and the resulting sine wave of current superimposed on the same time axis. Notice that as the voltage increases in a positive direction, the current increases along with it. When the voltage reverses direction, the current reverses direction. At all times, the voltage and current pass through the same relative parts of their respective cycles at the same time. When two waves, such as those shown in Figure 6.7, are precisely in step with one another, they are said to be *in phase.* To be in phase, the two waves must go through their maximum and minimum points at the same time and in the same direction.

In some circuits, several sine waves can be in phase with each other. Thus, it is possible to have two or more voltage drops in phase with each other and also in phase with the circuit current.

IMPORTANT

Note:** It is important to remember that Ohm's Law for d-c circuits is applicable to a-c circuits **with resistance only.

Voltage waves are not always in phase. For example, Figure 6.8 shows a voltage wave E_1 considered to start at 0° (time 1). As voltage wave E_1 reaches its positive peak, a second voltage wave, E_2, starts to rise (time 2). Because these waves do not go through their maximum and minimum points at the same instant of time, a **phase difference** exists between the two waves. The two waves are said to be out of phase. For the two waves in Figure 6.8, this phase difference is 90°.

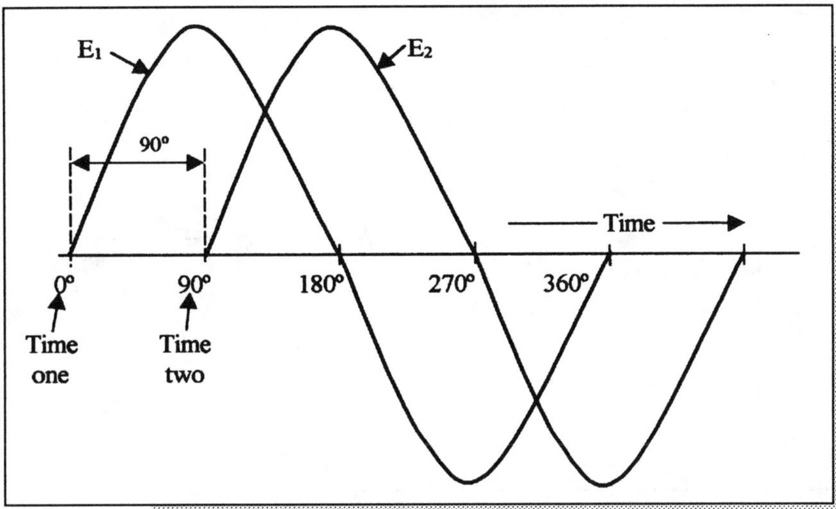

Figure 6.8
Voltage waves 90° out of phase.

6.4.1 PHASE RELATIONSHIPS

In the preceding section, we discussed the important concepts of **in phase** and **phase difference.** Another important phase concept is phase angle. The *phase angle* between two waveforms of the same frequency is the angular difference at a given instant of time. As an example, the phase angle between waves B and A (see Figure 6.9) is 90°. Take the instant of time at 90°. The horizontal axis is shown in angular units of time. Wave B starts at maximum value and reduces to zero value at 90°, while Wave A starts at zero and increases to maximum value at 90°. Wave B reaches its maximum value 90° ahead of wave A, so wave B **leads** wave A by 90° (and wave A **lags** wave B by 90°). This 90° phase angle between Waves B and A is maintained throughout the complete cycle and all successive cycles. At any instant of time, wave B has the value that wave A will have 90° later. Wave B is a cosine wave because it is displaced 90° from wave A, which is a sine wave.

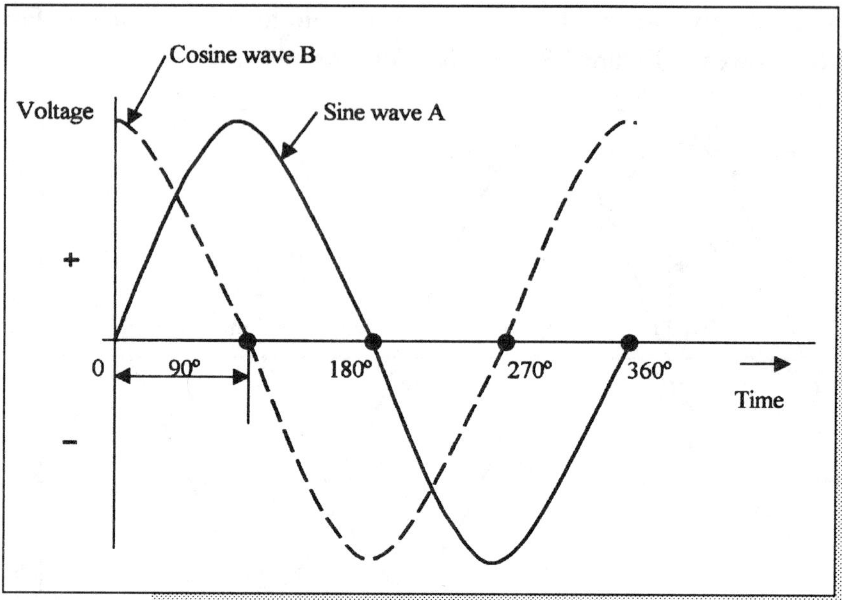

Figure 6.9
Wave B leads wave A by a phase angle of 90°.

IMPORTANT

Important Point: *The amount by which one wave leads or lags another is measured in degrees.*

To compare phase angles or phases of alternating voltages or currents, it is more convenient to use vector diagrams corresponding to the voltage and current waveforms. A *vector* is a straight line used to denote the magnitude and direction of a given quantity. Magnitude is denoted by the length of the line drawn to scale, and the direction is indicated by the arrow at one end of the line, together with the angle that the vector makes with a horizontal reference vector.

IMPORTANT

Note: *In electricity, because different directions really represent* ***time*** *expressed as a phase relationship, an electrical vector is called a* ***phasor.*** *In an a-c circuit containing only resistance, the voltage and current occur at the* ***same time,*** *or are in phase. To indicate this condition by means of phasors, all that is necessary is to draw the phasors for the voltage and current in the same direction. The value of each is indicated by the* ***length*** *of the phasor.*

A vector, or phasor, diagram is shown in Figure 6.10, where vector V_B is vertical to show the phase angle of 90° with respect to vector V_A, which is the reference. Because lead angles are shown in the counterclockwise direction from the reference vector, V_B leads V_A by 90°.

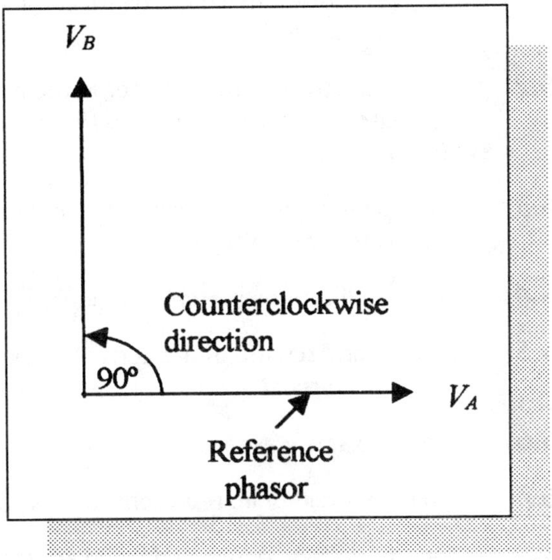

Figure 6.10
Phasor diagram.

Self-Test

6.1 As the armature of an alternator moves through one complete rotation, what must occur for maximum voltage to be generated?

6.2 If a conductor is rotated faster in the magnetic field of an alternator, what will happen to the frequency of the voltage?

6.3 If the peak a-c voltage across a resistor is 200 volts, what is the rms voltage?

6.4 What is electromagnetic induction?

6.5 What characteristic of a-c current distinguishes it from d-c current?

6.6 What is a sine wave?

6.7 What two values are represented by a vector?

6.8 The rms voltage generated by an alternator is 600 volts. What is the peak voltage?

6.9 What is frequency?

6.10 What are the two factors that affect the frequency of the voltage generated by an alternator?

Inductance

TOPICS

Inductance: What Is It?
Unit of Inductance
Self-Inductance
Mutual Inductance
Calculation of Total Inductance

7.1 INTRODUCTION

To this point we have learned the following key points about magnetic fields:

- A field of force exists around a wire carrying a current.
- This field has the form of concentric circles around the wire, in planes perpendicular to the wire, and with the wire at the center of the circles.
- The strength of the field depends on the current. Large currents produce large fields; small currents produce small fields.
- When lines of force cut across a conductor, a voltage is induced in the conductor.

Moreover, to this point, we have studied circuits that have been **resistive** (i.e., resistors presented the only opposition to current flow). Two other phenomena—inductance and capacitance—exist in d-c circuits to some extent, but they are major players in a-c circuits. Both inductance and capacitance present a kind of opposition to current flow that is called **reactance,** which we will study in Chapter 9. Before we examine reactance, however, we must first study inductance (the subject of this chapter) and capacitance (the subject of Chapter 8).

7.2 INDUCTANCE: WHAT IS IT?

Inductance is the characteristic of an electrical circuit that makes itself evident by opposing the starting, stopping, or changing of current

Key Terms Used in This Chapter

INDUCTANCE	A part of the total resistance of an electrical circuit, tending to resist the flow of changing current.
SELF-INDUCTANCE	The emf induced in a current-carrying conductor.
MUTUAL INDUCTANCE	Occurs when the magnetic field from one coil cuts through another coil.

flow. A simple analogy can be used to explain inductance. We are all familiar with how difficult it is to push a heavy load (a cart full of heavy materials, etc.). It takes more work to start the load moving than it does to keep it moving. This is because the load possesses the property of **inertia.** Inertia is the characteristic of mass that opposes a **change** in velocity. Therefore, inertia can hinder us in some ways and help us in others. Inductance exhibits the same effect on current in an electric circuit as inertia does on velocity of a mechanical object. The effects of inductance are sometimes desirable and sometimes undesirable.

IMPORTANT

Important Point: *Simply put,* ***inductance*** *is the characteristic of an electrical conductor that opposes a* ***change*** *in current flow.*

This means that because inductance is the property of an electric circuit that opposes any *change* in the current through that circuit, if the current increases, a self-induced voltage opposes this change and delays the increase. On the other hand, if the current decreases, a self-induced voltage tends to aid (or prolong) the current flow, delaying the decrease. Thus, current can neither increase nor decrease as fast in an inductive circuit as it can in a purely resistive circuit.

In a-c circuits, this effect becomes very important because it affects the **phase** relationships between voltage and current. We learned in Chapter 6 that voltages (or currents) can be out of phase if they are induced in separate armatures of an alternator. In that case, the voltage and current generated by each armature were in phase. When inductance is a factor in a circuit, the voltage and current generated by the **same** armature are out of phase. We shall examine these phase relationships later in this manual. Our objective in this chapter is to understand the nature and effects of inductance in an electric circuit.

7.3 UNIT OF INDUCTANCE

The unit for measuring inductance, L, is the *henry* (named for the American physicist, Joseph Henry), abbreviated h and normally written in lower case, henry. Figure 7.1 shows the schematic symbol for an inductor. An inductor has an inductance of 1 henry if an emf of 1 volt is induced in the inductor when the current through the inductor is changing at the rate of 1 ampere per second. The relation between the induced voltage, inductance, and rate of change of current with respect to time is stated mathematically as

$$E = L \frac{\Delta I}{\Delta t} \qquad (7.1)$$

where

E = the induced emf in volts
L = the inductance in henrys
ΔI = is the change in amperes occurring in Δt seconds

Note: *The symbol Δ (delta) means "a change in . . . "*

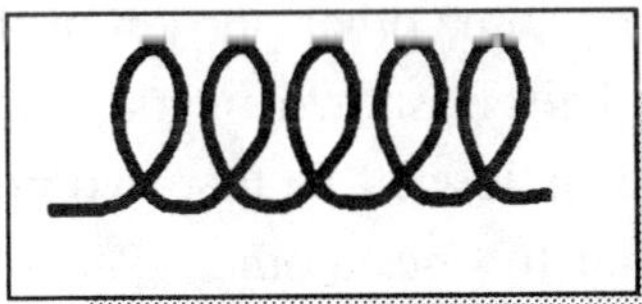

Figure 7.1
Schematic symbol for an inductor.

The henry is a large unit of inductance and is used with relatively large inductors. The unit employed with small inductors is the millihenry (mh). For still smaller inductors, the unit of inductance is the microhenry (μh).

7.4 SELF-INDUCTANCE

As previously explained, current flow in a conductor always produces a magnetic field surrounding, or linking with, the conductor. When the current changes, the magnetic field changes, and an emf is induced in the conductor. This emf is called a *self-induced emf* because it is induced in the conductor carrying the current.

IMPORTANT

Note: *Even a perfectly straight length of conductor has some inductance.*

The direction of the induced emf has a definite relation to the direction in which the field that induces the emf varies. When the current in a circuit is increasing, the flux linking with the circuit is increasing. This flux cuts across the conductor and induces an emf in the conductor in such a direction to oppose the increase in current and flux. This emf is sometimes referred to as **counterelectromotive force** (cemf). The two terms are used synonymously throughout this manual. Likewise, when the current is decreasing, an emf is induced in the opposite direction and opposes the decrease in current.

IMPORTANT

Important Point: *The effects just described are summarized by* ***Lenz's Law,*** *which states that the induced emf in any circuit is always in a direction opposed to the effect that produced it.*

Shaping a conductor so that the electromagnetic field around each portion of the conductor cuts across some other portion of the same conductor increases the inductance. This is shown in its simplest form in Figure 7.2(a). A loop of conductor is looped so that two portions of the conductor lie adjacent and parallel to one another. These portions are labeled Conductor 1 and Conductor 2. When the switch is closed, electron flow through the conductor establishes a typical concentric field around **all** portions of the conductor. The field is shown in a single plane (for simplicity) that is perpendicular to both conductors. Although the field originates simultaneously in both conductors, it is considered as originating in Conductor 1, and its effect on Conductor 2 will be noted.

With increasing current, the field expands outward, cutting across a portion of Conductor 2. The resultant induced emf in Conductor 2 is shown by the dashed arrow. Note that it is in **opposition** to the battery current and voltage, according to Lenz's Law.

In Figure 7.2(b), the same section of Conductor 2 is shown, but with the switch opened and the flux collapsing.

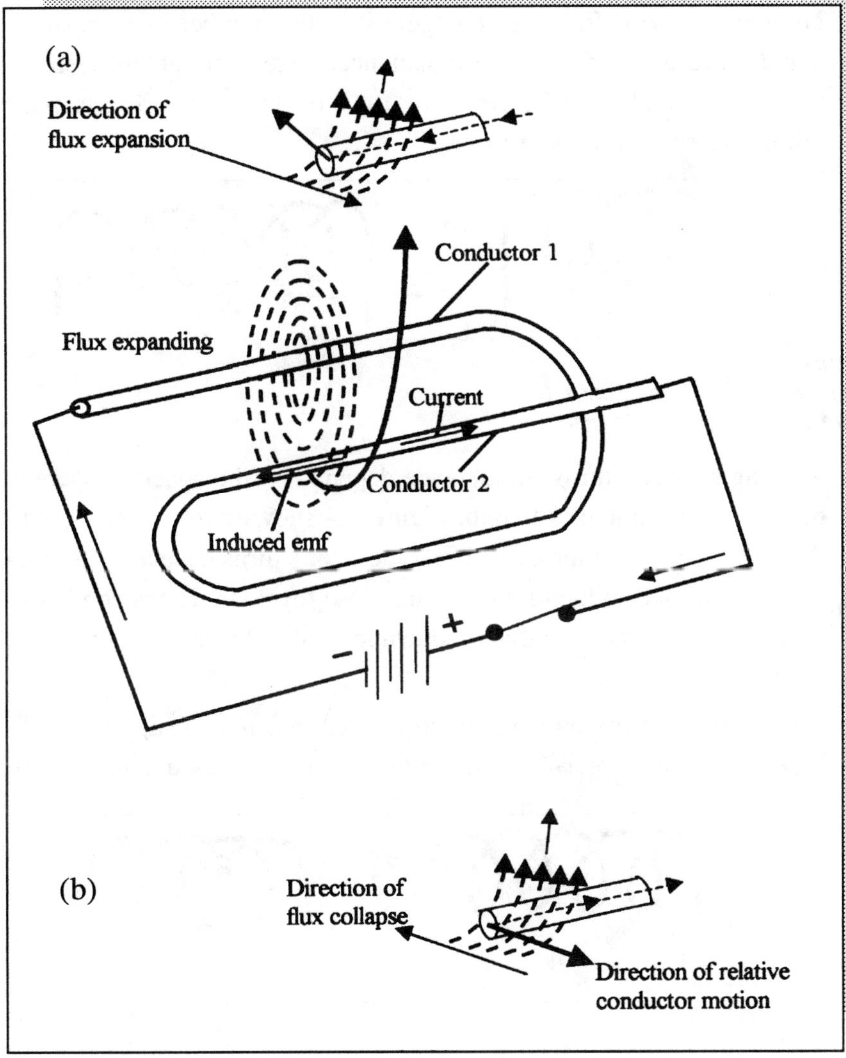

Figure 7.2
Self-inductance.

IMPORTANT ***Important Point:*** *From Figure 7.2, the important point to note is that the voltage of self-induction opposes both* changes *in current. It delays the initial buildup of current by opposing the battery voltage and delays the breakdown of current by exerting an induced voltage in the same direction that the battery voltage acted.*

Four major factors affect the self-inductance of a conductor, or circuit.

1. Number of turns: Inductance depends on the number of wire turns. Wind more turns to increase inductance. Take turns off to decrease the inductance. Figure 7.3 compares the inductance of two coils made with different numbers of turns.

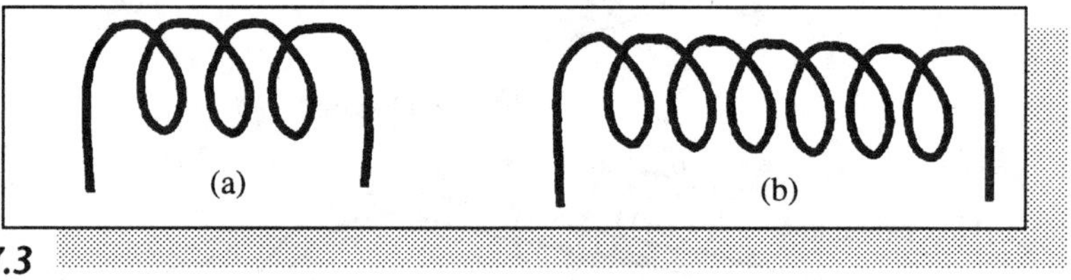

Figure 7.3
(a) Few turns, low inductance; (b) more turns, higher inductance.

2. Spacing between turns: Inductance depends on the spacing between turns, or the inductor's length. Figure 7.4 shows two inductors with the same number of turns. The first inductor's turns have a wide spacing. The second inductor's turns are close together. The second coil, though shorter, has a larger inductance value because of its close spacing between turns.
3. Coil diameter: Coil diameter, or cross-sectional area, is highlighted in Figure 7.5. The larger-diameter inductor has more inductance. Both

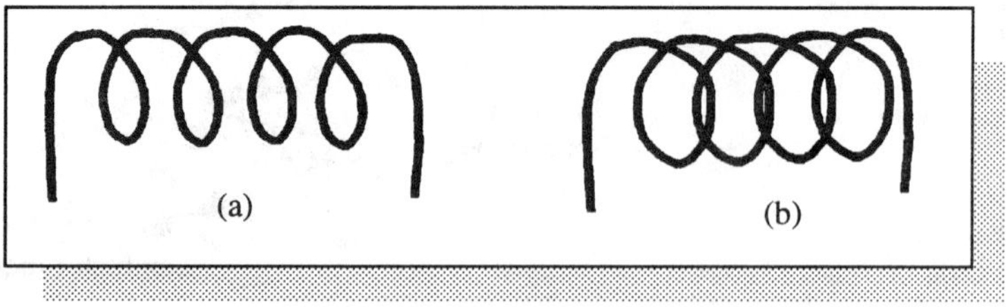

Figure 7.4
(a) Wide spacing between turns, low inductance; (b) close spacing between turns, higher inductance.

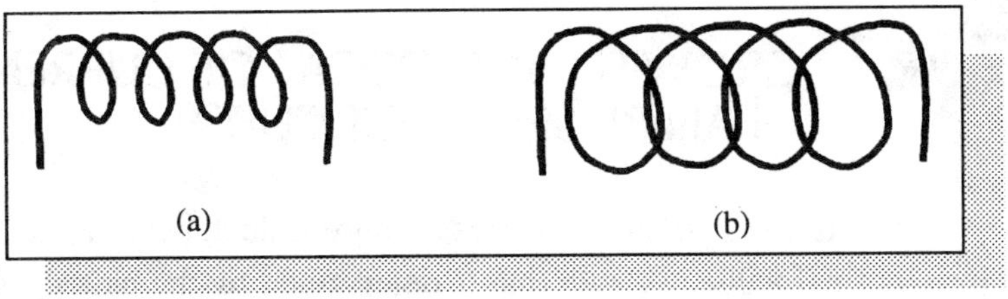

Figure 7.5
(a) Small diameter, low inductance; (b) larger diameter, higher inductance.

coils shown have the same number of turns, and the spacing between turns is the same. The first inductor has a small diameter, and the second one has a larger diameter. The second inductor has more inductance than the first one.

4. Type of core material: **Permeability,** as pointed out earlier, is a measure of how easily a magnetic field goes through a material. Permeability also tells us how much stronger the magnetic field will be with the material inside the coil.

   Figure 7.6 shows three identical coils. One has an air core, one has a powdered-iron core in the center, and the other has a soft-iron core. This figure illustrates the effects of core material on inductance. The inductance of a coil is affected by the magnitude of current when the core is a magnetic material. When the core is air, the inductance is independent of the current.

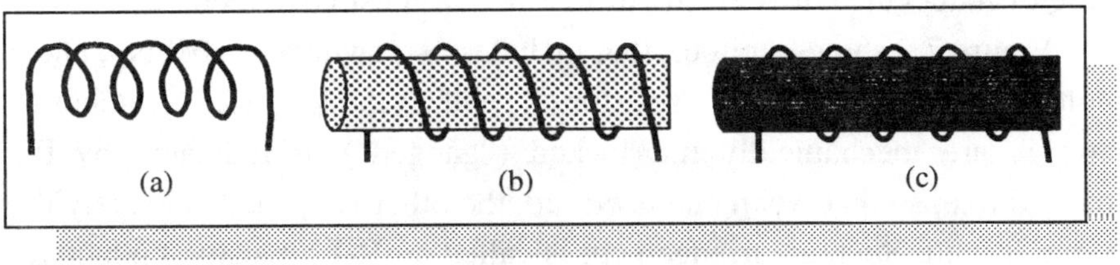

Figure 7.6
(a) Air core, low inductance; (b) powdered-iron core, higher inductance; (c) soft- iron core, highest inductance.

IMPORTANT ***Key Point:*** *The inductance of a coil increases very rapidly as the number of turns is increased. It also increases as the coil is made shorter, the cross-sectional area is made larger, or the permeability of the core is increased.*

7.4.1 GROWTH AND DECAY OF CURRENT IN AN RL SERIES CIRCUIT

If a battery is connected across a pure inductance, the current builds up to its final value at a rate that is determined by the battery voltage and the internal resistance of the battery. The current buildup is gradual because of the counter emf (cemf) generated by the self-inductance of the coil. When the current starts to flow, the magnetic lines of force move out, cut the turns of wire on the inductor, and build up a cemf that opposes the emf of the battery. This opposition causes a delay in the time it takes the current to build up to steady value. When the battery is disconnected, the lines of force collapse, again cutting the turns of the inductor and building up an emf that tends to prolong the current flow.

Although the analogy is not exact, electrical inductance is somewhat like mechanical inertia. A boat begins to move on the surface of water at the instant a constant force is applied to it. At this instant, its rate of change of speed (acceleration) is greatest, and all the applied force is used to overcome the inertia of the boat. After a while, the speed of the boat increases (its acceleration decreases), and the applied force is used up in overcoming the friction of the water against the hull. As the speed levels off and the acceleration becomes zero, the applied force equals the opposing friction force at this speed, and the inertia effect disappears. In the case of inductance, it is electrical inertia that must be overcome.

Figure 7.7 shows a circuit that includes two switches, a battery, and a voltage divider containing a resistor (R) and an inductor (L). Switches S_1 and S_2 are mechanically interlocked ("ganged") as indicated by the dashed line, so that when one is closed, the other is opened at exactly the same instant. Such an arrangement is called an **R-L** (resistive-inductive) series circuit. The source voltage of the battery is applied across the R-L combination (the resistor and inductor) when S_1 is closed. As S_1 is closed, as shown in Figure 7.7, a voltage (E) appears across the R-L combination. A current attempts to flow, but the inductor opposes this current by building up a cemf. At the first instant S_1 is closed, the cemf *exactly equals* the battery emf, and its polarity is opposite. Under this condition, no current will flow in resistor R. Because no current can flow when the cemf is exactly equal to the battery voltage, no voltage is dropped across R. As

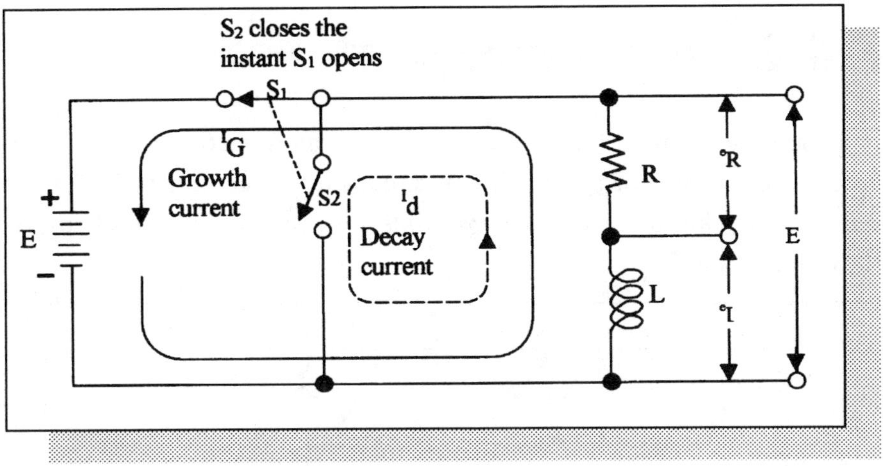

Figure 7.7
Growth and decay of current in an RL circuit.

time goes on, more of the battery voltage appears across the resistor and less across the inductor. The rate of change of current is less, and the induced emf is less. As the steady-state condition of the current flow is approached, the drop across the inductor approaches zero, and all the battery voltage is used to overcome the resistance of the circuit.

When S_2 is closed (source voltage E removed from the circuit), the flux that has been established around L collapses through the windings and induces a voltage, e_L, in L that has a polarity opposite to E and essentially equal to it in magnitude. The induced voltage, e_L, causes i_d to flow through R in the same direction that it was flowing when S_1 was closed. A voltage, e_R, that is initially equal to *E*, is developed across R. It rapidly falls to zero as the voltage, e_L, across L, due to the collapsing flux, falls to zero.

IMPORTANT

Key Point: *When a switch is first closed to complete a circuit, inductance opposes the buildup of current in the circuit.*

IMPORTANT

Note: *In many electronic circuits, the time required for the growth or decay of current is important. However, these applications are beyond the scope of this manual, but you need to learn the fundamentals of the L/R time constant, which are discussed in the following section.*

7.4.2 *L/R* TIME CONSTANT

The time required for the current through an inductor to increase to 63.2 percent (63%) of the maximum current or to decrease to 36.7 percent (37%) is known as the *time constant* of the circuit. An RL circuit is shown in Figure 7.8.

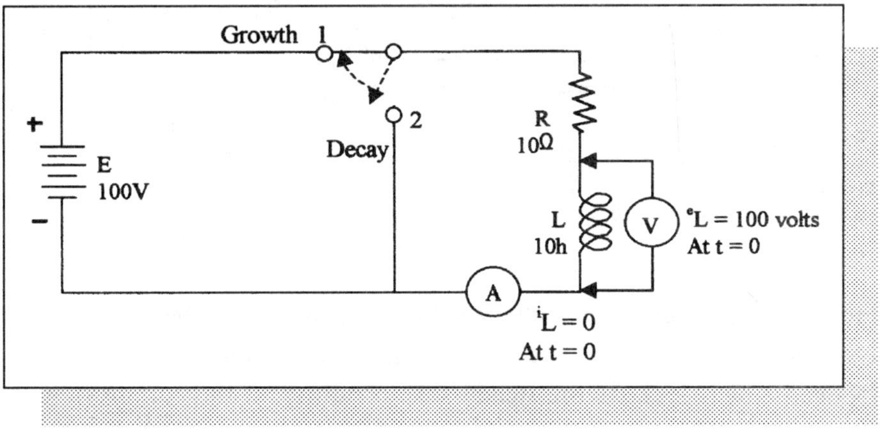

Figure 7.8
L/R *time constant.*

The value of the time constant in seconds is equal to the inductance in henrys divided by the circuit resistance in ohms. One set of values is given in Figure 7.8. *L/R* is the symbol used for this time constant. If *L* is in henrys and *R* is in ohms, *t* (time) is in seconds. If *L* is in microhenrys and *R* is in ohms, *t* is in microseconds. If *L* is in millihenrys and *R* is in ohms, *t* is in milliseconds.

R in the *L/R* equation is always in ohms, and the time constant is on the same order of magnitude as *L*. Two useful relations used in calculating *L/R* time constants are as follows:

$$\frac{L \text{ (in henrys)}}{R \text{ (in ohms)}} = t \text{ (in seconds)} \qquad (7.2)$$

$$\frac{L \text{ (microhenrys)}}{R \text{ (in ohms)}} = t \text{ (in milliseconds)} \qquad (7.3)$$

Key Point: *The time constant of an L/R circuit is always expressed as a ratio between inductance (or L) and resistance (or R).*

IMPORTANT

7.5 MUTUAL INDUCTANCE

When the current in a conductor or coil changes, the varying flux can cut across any other conductor or coil located nearby, thus inducing voltages in both. A varying current in L_1, therefore, induces voltage across L_1 and across L_2 (see Figure 7.9; see Figure 7.10 for the schematic symbol for two coils with mutual inductance). When the induced voltage e_{L2} produces current in L_2, its varying magnetic field induces voltage in L_1. Hence, the two coils L_1 and L_2 have *mutual inductance* because current change in one coil can induce voltage in the other. The unit of mutual inductance is the henry, and the symbol is L_M. Two coils have L_M of 1 H when a current change of 1 A/s in one coil induces 1 E in the other coil.

The factors affecting the mutual inductance of two adjacent coils is dependent upon

- physical dimensions of the two coils
- number of turns in each coil
- distance between the two coils
- relative positions of the axes of the two coils
- the permeability of the cores

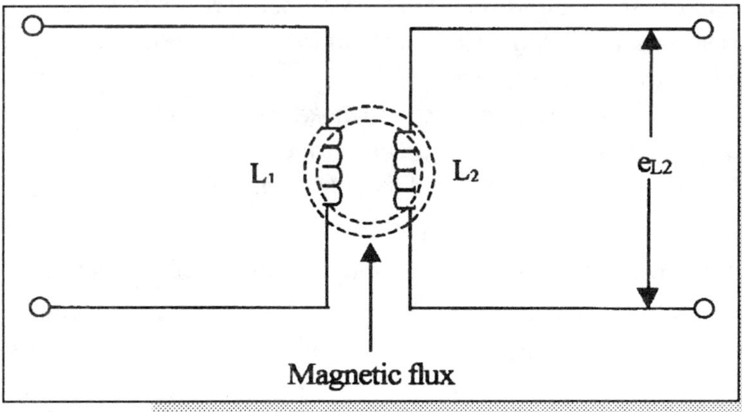

Figure 7.9
Mutual inductance between L_1 and L_2.

Figure 7.10
(a) Schematic symbol for two coils (air core) with mutual inductance; (b) two coils (iron core) with mutual inductance.

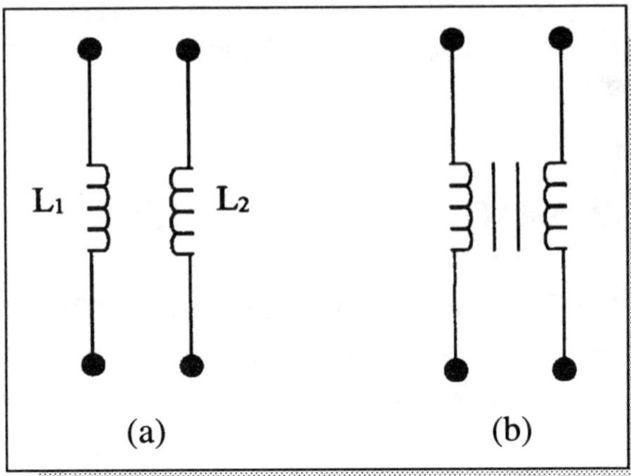

IMPORTANT

Important Point: *The amount of mutual inductance depends on the relative position of the two coils. If the coils are separated a considerable distance, the amount of flux common to both coils is small, and the mutual inductance is low. Conversely, if the coils are close together so that nearly all the flow of one coil links the turns of the other, mutual inductance is high. The mutual inductance can be increased greatly by mounting the coils on a common iron core.*

7.6 CALCULATION OF TOTAL INDUCTANCE

IMPORTANT

Note: *In the study of advanced electrical theory, it is necessary to know the effect of mutual inductance in solving for total inductance in both series and parallel circuits. However, for our purposes, in this manual, we do not attempt to make these calculations. Instead, we discuss the basic total inductance calculations with which the maintenance operator should be familiar.*

If inductors in series are located far enough apart, or are well shielded to make the effects of mutual inductance negligible, the total inductance is calculated in the same manner as for resistances in series; we merely add them:

$$L_T = L_1 + L_2 + L_3 \ldots. \text{(etc.)} \qquad (7.4)$$

EXAMPLE 7.1

Problem: If a series circuit contains three inductors whose values are 40 µh, 50 µh, and 20 µh, what is the total inductance?

Solution:

$$L_T = 40\mu h + 50\mu h + 20\mu h$$
$$= 110\ \mu h$$

In a parallel circuit containing inductors (without mutual inductance), the total inductance is calculated in the same manner as for resistances in parallel:

$$\frac{1}{L_T} = \frac{1}{L_1} + \frac{1}{L_2} + \frac{1}{L_3} + \ldots \text{(etc.)} \qquad (7.5)$$

EXAMPLE 7.2

Problem: A circuit contains three totally shielded inductors in parallel. The values of the three inductances are 4 mh, 5 mh, and 10 mh. What is the total inductance?

Solution:

$$\frac{1}{L_T} = \frac{1}{4} + \frac{1}{5} + \frac{1}{10}$$

$$= 0.25 + 0.2 + 0.1$$

$$= 0.55$$

$$L_T = \frac{1}{0.55}$$

$$= 1.8\ \text{mh}$$

Self-Test

7.1 When an a-c voltage is impressed across a coil, the resulting current is an ________ current. This changing current produces changing fields of force that ________ the wires of the coil. These cuttings induce a __________ emf in the coil.

7.2 The inductance of a coil is a measure of its ability to produce a ___________ emf when the current through it is changing.

7.3 What characteristics of electric current result in self-inductance?

7.4 How is a self-induced emf produced in the conductor?

7.5 List four factors affecting the inductance of a coil.

7.6 An increase in the number of turns in a coil will ____________ inductance.

7.7 An increase in the length of a coil will ___________ inductance.

7.8 The inductance of an R/L circuit is 20 mh, and the resistance is 5 ohms. What is the L/R time constant of the circuit?

7.9 An increase in the cross-sectional area of a coil will______________ inductance.

7.10 An increase in the permeability of a coil will ______________ inductance.

Capacitance

TOPICS

Capacitance: What Is It?
The Capacitor
Dielectric Materials
Unit of Capacitance
Factors Affecting the Value of Capacitance
Voltage Rating of Capacitors
Charge and Discharge of an RC Series Circuit
RC Time Constant
Capacitors in Series and in Parallel
Types of Capacitors

8.1 INTRODUCTION

No matter how complex the electrical circuit, it is composed of no more than three basic electrical properties: resistance, inductance, and capacitance. Accordingly, gaining a thorough understanding of each of these three basic properties is a necessary step toward the understanding of electrical equipment. Because resistance and inductance have been covered, the last of the basic three, capacitance, is covered in this chapter.

8.2 CAPACITANCE: WHAT IS IT?

In Chapter 7, we learned that inductance opposes any change in current. *Capacitance* is the property of an electric circuit that opposes any change of *voltage* in a circuit. That is, if applied voltage is increased, capacitance opposes the change and delays the voltage increase across the circuit. If applied voltage is decreased, capacitance tends to maintain the higher original voltage across the circuit, thus delaying the decrease.

Capacitance is also defined as that property of a circuit that enables energy to be stored in an electric field. Natural capacitance exists in many

Key Terms Used in This Chapter

CAPACITOR	Is made of two metallic plates separated by an insulating material.
CAPACITANCE	A measure of a capacitor's ability to store electrons
WORKING VOLTAGE	The largest voltage that may be placed across a capacitor before it breaks down.
FARAD	The capacitance of a capacitor that can hold 1 coulomb of electrons on its plates under a pressure of 1 volt.

electric circuits. However, in this manual, we are concerned only with the capacitance that is designed into the circuit by means of devices called *capacitors.*

IMPORTANT **Key Point:** *The most noticeable effect of capacitance in a circuit is that voltage can neither increase nor decrease as rapidly in a capacitive circuit as it can in a circuit that does not include capacitance.*

8.3 THE CAPACITOR

A *capacitor,* or condenser, is a manufactured electrical device that consists of two conducting plates of metal separated by an insulating material called a *dielectric* (see Figure 8.1). (Note: the prefix "di" means "through" or "across.")

The schematic symbol for a capacitor is shown in Figure 8.2.

When a capacitor is connected to a voltage source, there is a short current pulse. A capacitor stores this electric charge in the dielectric (it can be charged and discharged, as we shall see later). To form a capacitor of any appreciable value, however, the area of the metal pieces must be quite large, and the thickness of the dielectric must be quite small.

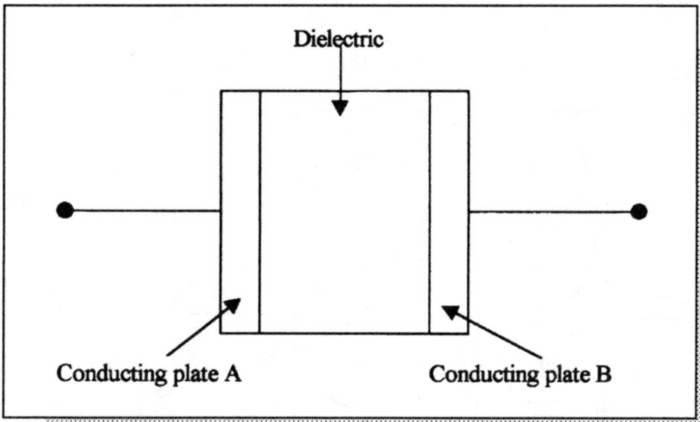

Figure 8.1
Capacitor.

IMPORTANT

Key Point: *A capacitor is essentially a device that stores electrical energy.*

The capacitor is used in a number of ways in electrical circuits. It may block d-c portions of a circuit, because it is effectively a barrier to direct current (but not to a-c current). It may be part of a tuned circuit—one such application is in the tuning of a radio to a particular station. It may be used to filter a-c out of a d-c circuit. Most of these are advanced applications that are beyond the scope of this manual, however, a basic understanding of capacitance is necessary to the fundamentals of a-c theory.

IMPORTANT

Important Point: *A capacitor does not conduct d-c current. The insulation between the capacitor plates blocks the flow of electrons. We learned earlier that there is a short current pulse when we first connect the capacitor to a voltage source. The capacitor quickly charges to the supply voltage, and then the current stops.*

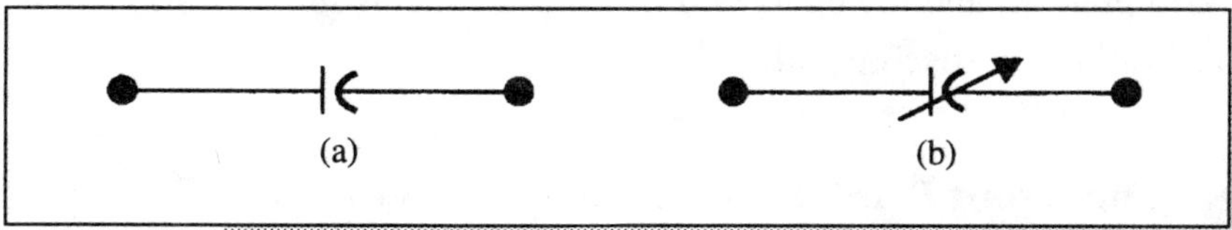

Figure 8.2
(a) Schematic for a fixed capacitor;
(b) variable capacitor.

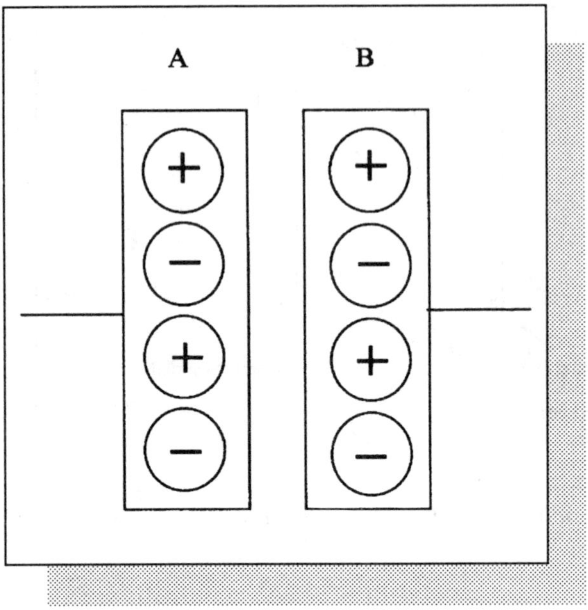

Figure 8.3
Two plates of a capacitor with a neutral charge.

The two plates of the capacitor shown in Figure 8.3 are electrically neutral because there are as many protons (positive charge) as electrons (negative charge) on each plate. Thus, the capacitor has **no charge.**

Now a battery is connected across the plates [see Figure 8.4(a)]. When the switch is closed [see Figure 8.4(b)], the negative charge on plate A is attracted to the positive terminal of the battery. This movement of charges will continue until the difference in charge between plates A and B is equal to the electromotive force (voltage) of the battery.

The capacitor is now **charged.** Because almost none of the charge can cross the space between plates, the capacitor will remain in this condition even if the battery is removed [see Figure 8.5(a)]. However, if a conductor is placed across the plates [see Figure 8.5(b)], the electrons find a path back to plate A, and the charges on each plate are again neutralized. The capacitor is now **discharged.**

IMPORTANT ***Important Point:*** *In a capacitor, electrons cannot flow through the dielectric, because it is an insulator. Because it takes a definite quantity of electrons to charge (fill up) a capacitor, it is said to have* ***capacity.*** *This characteristic is referred to as* capacitance.

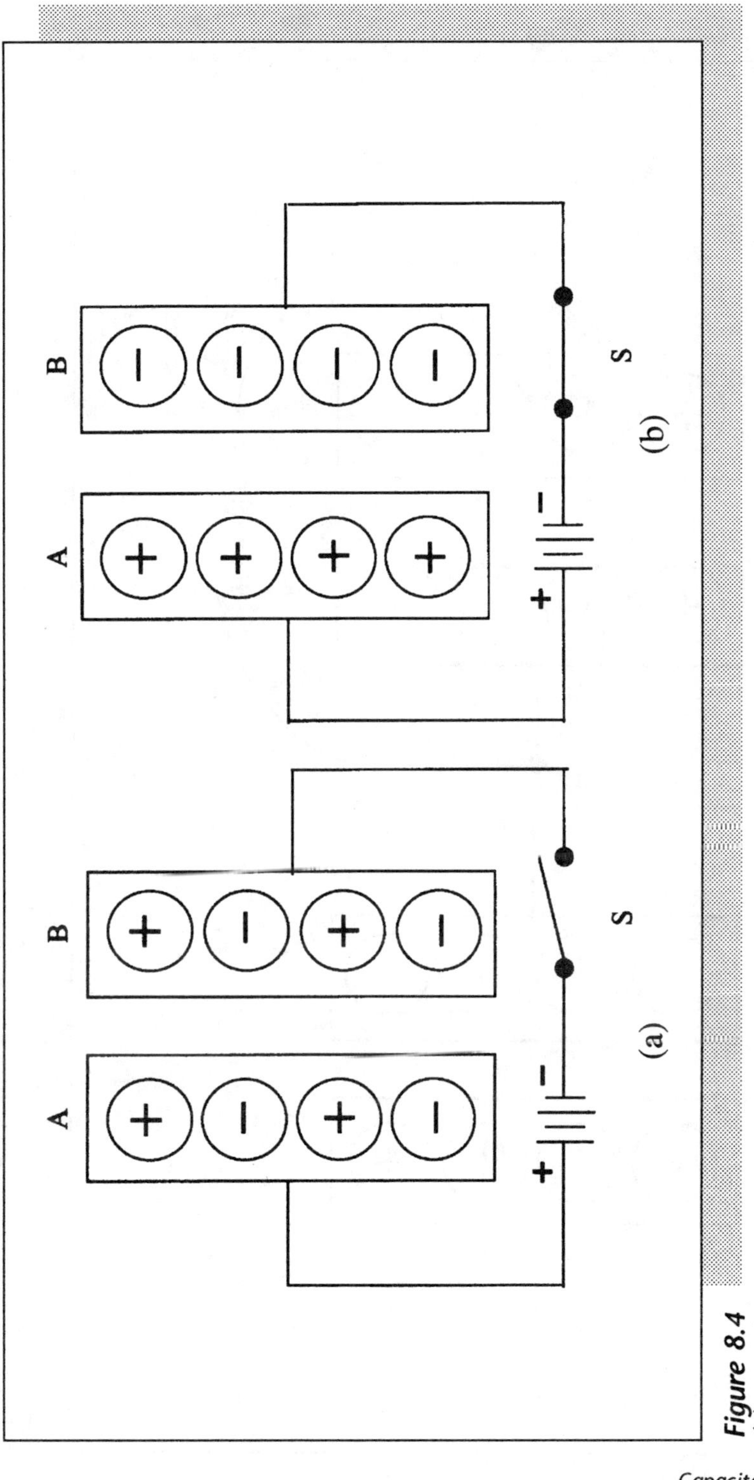

Figure 8.4
(a) Neutral capacitor;
(b) charged capacitor.

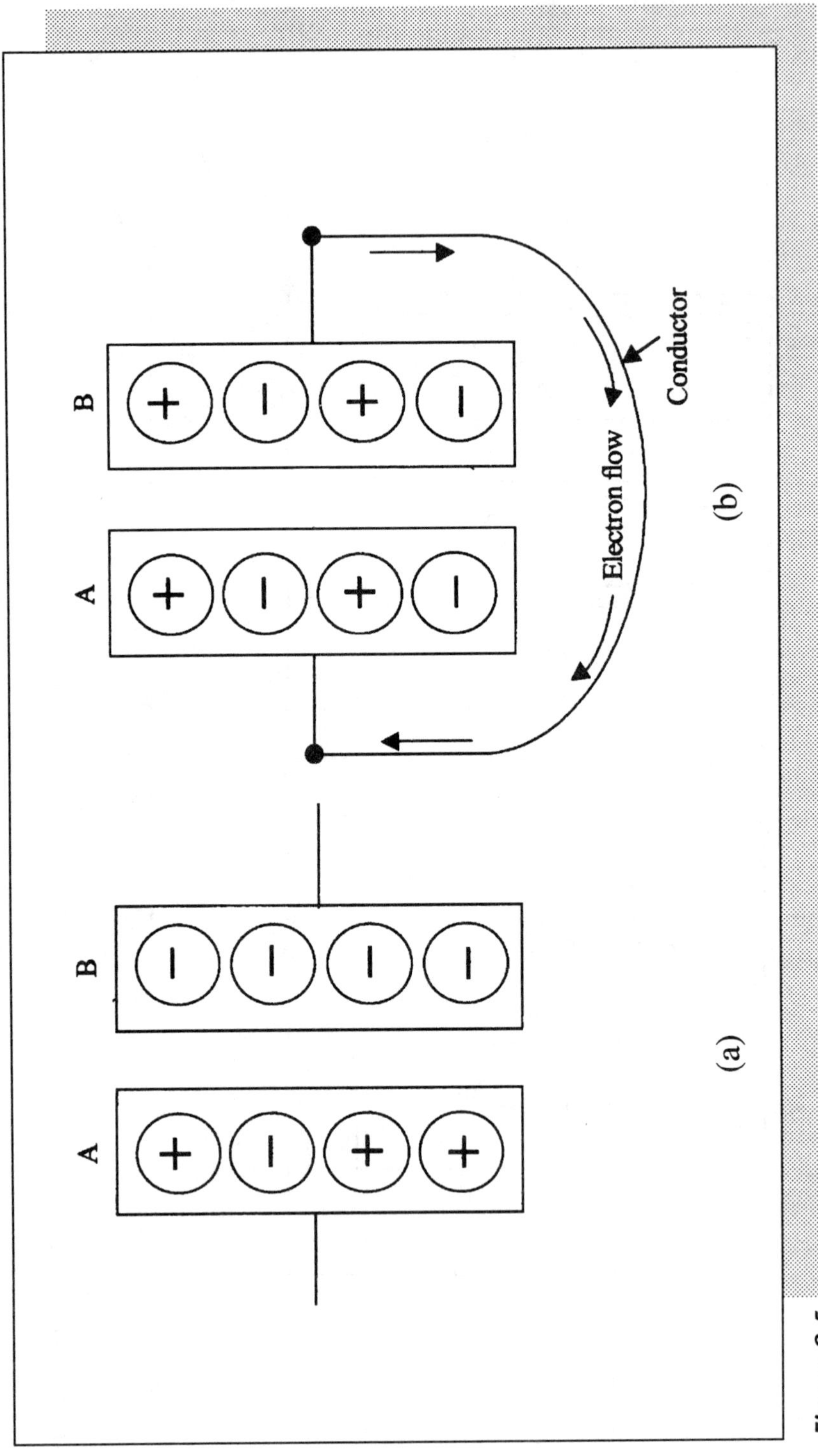

Figure 8.5
(a) Charged capacitor; (b) discharging a capacitor.

8.4 DIELECTRIC MATERIALS

Somewhat similar to the phenomenon of permeability in magnetic circuits, various materials differ in their ability to support electric flux (lines of force) or to serve as dielectric material for capacitors. Materials are rated in their ability to support electric flux in terms of a number called a *dielectric constant*. Other factors being equal, the higher the value of the dielectric constant, the better is the dielectric material. Dry air is the standard (the reference) by which other materials are rated. Dielectric constants for some common materials are given in Table 8.1.

TABLE 8.1. Dielectric Constants.

Material	Constant
Vacuum	1.0000
Air	1.0006
Paraffin paper	3.5
Glass	5–10
Quartz	3.8
Mica	3–6
Rubber	2.5–35
Wood	2.5–8
Porcelain	5.1–5.9
Glycerine (15°C)	56
Petroleum	2
Pure water	81

IMPORTANT

Note: *From Table 8.1, it is obvious that "pure" water is the best dielectric. Keep in mind that the key word is "pure." Water capacitors are used today in some high-energy applications, in which differences in potential are measured in thousands of volts.*

8.5 UNIT OF CAPACITANCE

Capacitance is equal to the amount of charge that can be stored in a capacitor divided by the voltage applied across the plates

$$C = \frac{Q}{E} \tag{8.1}$$

where

C = capacitance, F (farads)
Q = amount of charge, C (coulombs)
E = voltage, V

EXAMPLE 8.1

Problem: What is the capacitance of two metal plates separated by 1 cm of air, if 0.002 coulomb of charge is stored when a potential of 300 volts is applied to the capacitor?

Solution:

Given: $Q = 0.001$ coulomb
$E = 200$ volts

$$C = \frac{Q}{E}$$

Converting to powers of 10

$$C = \frac{10 \times 10^{-4}}{2 \times 10^{2}}$$

$$C = 5 \times 10^{-6}$$

$$C = 0.000005 \text{ farads}$$

IMPORTANT

Note: *Although the capacitance value obtained in Example 8.1 appears small, many electronic circuits require capacitors of much smaller value. Consequently, the farad is a cumbersome unit, far too large for most applications. The* ***microfarad,*** *which is one millionth of a farad (*1×10^{-6} *farad), is a more convenient unit. The symbols used to designated microfarad are* μf.

Equation (8.1) can be rewritten as follows:

$$Q = CE \tag{8.2}$$

$$E = \frac{Q}{C} \tag{8.3}$$

IMPORTANT ***Important Point:*** *From Equation (8.2), do not deduce the mistaken idea that capacitance is dependent upon charge and voltage. Capacitance is determined entirely by physical factors, which are covered in Section 8.6.*

The symbol used to designate a capacitor is (C). The unit of capacitance is the farad (F). The farad is that capacitance that will store one coulomb of charge in the dielectric when the voltage applied across the capacitor terminals is one volt.

8.6 FACTORS AFFECTING THE VALUE OF CAPACITANCE

The capacitance of a capacitor depends on three main factors: plate surface area, distance between plates, and dielectric constant of the insulating material.

- **Plate surface area**—capacitance varies directly with plate surface area. We can double the capacitance value by doubling the capacitor's plate surface area. Figure 8.6 shows a capacitor with a small surface area and another one with a large surface area.

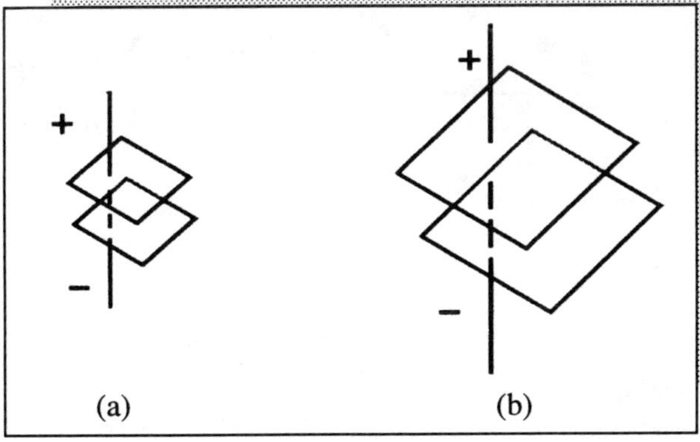

Figure 8.6
(a) Small plates, small capacitance; (a) larger plates, higher capacitance.

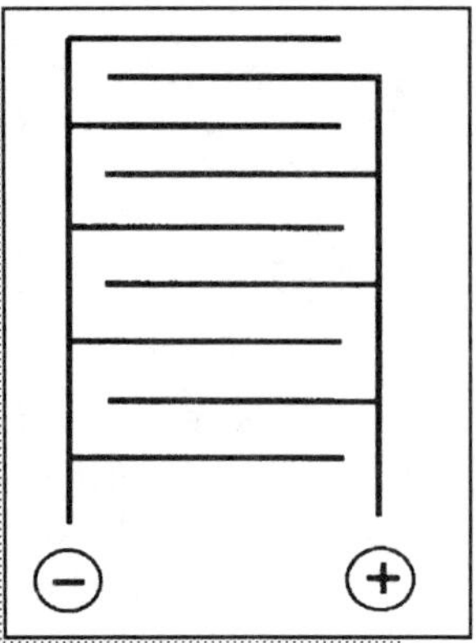

Figure 8.7 *Shows several sets of plates connected to produce a capacitor with more surface area.*

Adding more capacitor plates can increase the plate surface area. Figure 8.7 shows alternate plates connecting to opposite capacitor terminals.

- **Distance between plates**—capacitance varies inversely with the distance between plate surfaces. The capacitance increases when the plates are closer together. Figure 8.8 shows capacitors with the same plate surface area, but with different spacing.
- **Dielectric constant of the insulating material**—an insulating material with a higher dielectric constant produces a higher capacitance rating. Figure 8.9 shows two capacitors. Both have the same plate surface area and spacing. Air is the dielectric in the first capacitor, and mica is the dielectric in the second one. Mica's dielectric constant is 5.4 times greater than air's dielectric constant. The mica capacitor has 5.4 times more capacitance than the air-dielectric capacitor.

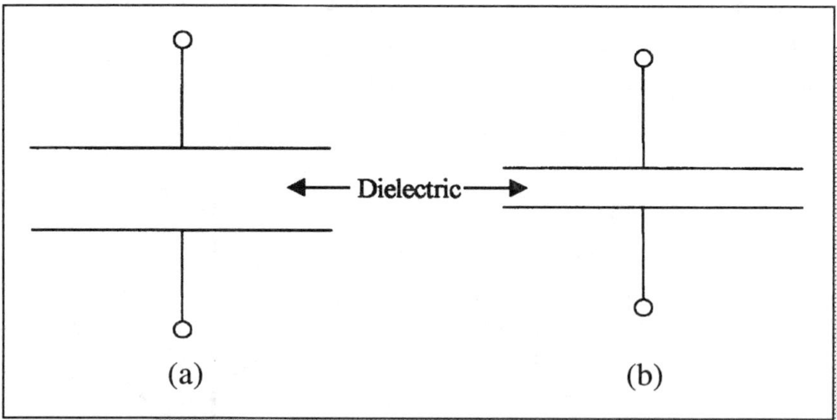

Figure 8.8
(a) Wide plate spacing, small capacitance;
(b) narrow plate spacing, larger capacitance.

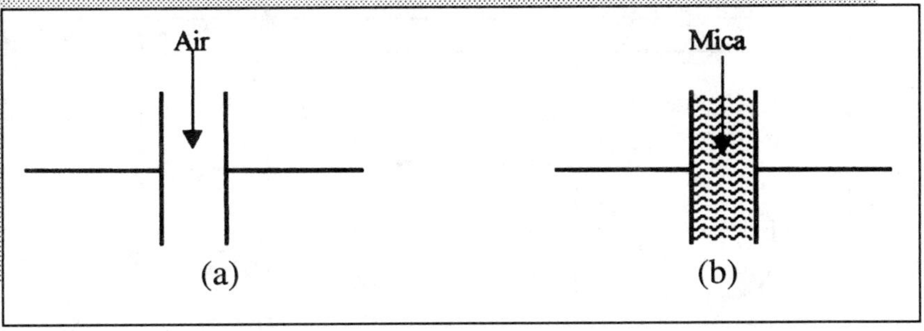

Figure 8.9
(a) Low capacitance;
(b) higher capacitance.

8.7 VOLTAGE RATING OF CAPACITORS

There is a limit to the voltage that may be applied across any capacitor. If too large a voltage is applied, it will overcome the resistance of the dielectric, and a current will be forced through it from one plate to the other, sometimes burning a hole in the dielectric. In this event, a short circuit exists, and the capacitor must be discarded. The maximum voltage that may be applied to a capacitor is known as the **working voltage** and must never be exceeded.

The working voltage of a capacitor depends on (1) the type of material used as the dielectric and (2) the thickness of the dielectric. As a margin of safety, the capacitor should be selected so that its working voltage is at least 50 percent greater than the highest voltage to be applied to it. For example, if a capacitor is expected to have a maximum of 200 volts applied to it, its working voltage should be at least 300 volts.

8.8 CHARGE AND DISCHARGE OF AN RC SERIES CIRCUIT

According to Ohm's Law, the voltage across a resistance is equal to the current through it times the value of the resistance. This means that a voltage will be developed across a resistance **only when current flows through it.**

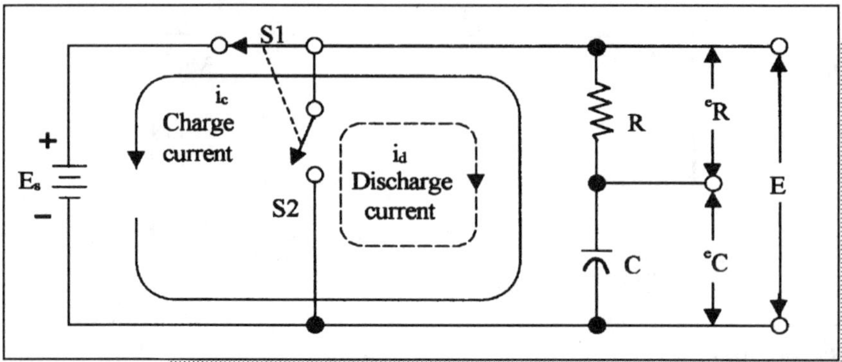

Figure 8.10
Charge and discharge of an RC series circuit.

As previously stated, a capacitor is capable of storing or holding a charge of electrons. When uncharged, both plates contain the same number of free electrons. When charged, one plate contains more free electrons than the other. The difference in the number of electrons is a measure of the charge on the capacitor. The accumulation of this charge builds up a voltage across the terminals of the capacitor, and the charge continues to increase until this voltage equals the applied voltage. The greater the voltage, the greater the charge on the capacitor. Unless a discharge path is provided, a capacitor keeps its charge indefinitely. Any practical capacitor, however, has some leakage through the dielectric so that the voltage will gradually leak off.

A voltage divider containing resistance and capacitance may be connected in a circuit by means of a switch, as shown in Figure 8.10. Such a series arrangement is called an **RC series circuit.**

If S1 is closed, electrons flow counterclockwise around the circuit containing the battery, capacitor, and resistor. This flow of electrons ceases when C is charged to the battery voltage. At the instant current begins to flow, there is no voltage on the capacitor, and the drop across R is equal to the battery voltage. The initial charging current, I, is therefore equal to E_s/R.

The current flowing in the circuit soon charges the capacitor. Because the voltage on the capacitor is proportional to its charge, a voltage, e_c, will appear across the capacitor. This voltage opposes the battery voltage—that is, these two voltages buck each other. As a result, the voltage e_R across the resistor is $E_S - e_C$, and this is equal to the voltage drop ($i_C R$) across the resistor. Because E_S is fixed, i_c decreases as e_c increases.

The charging process continues until the capacitor is fully charged and the voltage across it is equal to the battery voltage. At this instant, the voltage across is zero, and no current flows through it.

If S2 is closed (S1 opened) in Figure 8.10, a discharge current, i_d, will discharge the capacitor. Because i_d is opposite in direction to i_c, the voltage across the resistor will have a polarity opposite to the polarity during the charging time. However, this voltage will have the same magnitude and will vary in the same manner. During discharge, the voltage across the capacitor is equal and opposite to the drop across the resistor. The voltage drops rapidly from its initial value and then approaches zero slowly.

The actual time it takes to charge or discharge is important in advanced electricity and electronics. Because the charge or discharge time depends on the values of resistance and capacitance, an RC circuit can be designed for the proper timing of certain electrical events. RC time constant is covered in the next section.

8.9 RC TIME CONSTANT

The time required to charge a capacitor to 63% of maximum voltage or to discharge it to 37% of its final voltage is known as the *time constant* of the current. An RC circuit is shown in Figure 8.11.

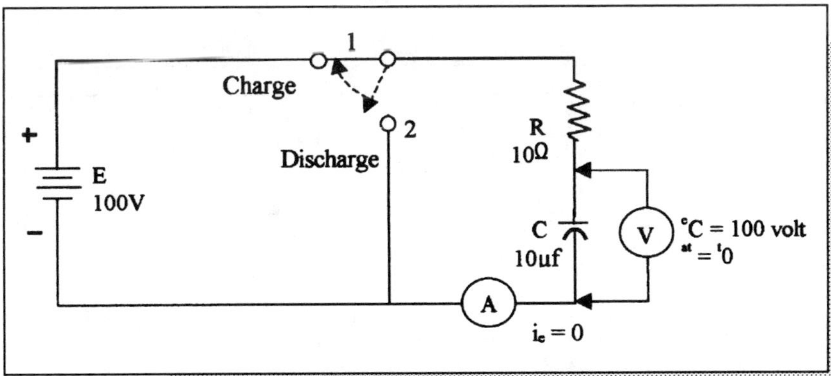

Figure 8.11
An RC circuit.

The time constant T for an RC circuit is

$$T = RC \tag{8.4}$$

The time constant of an RC circuit is usually very short because the capacitance of a circuit may be only a few microfarads or even picofarads.

IMPORTANT

Key Point: *An RC time constant expresses the charge and discharge times for a capacitor.*

8.10 CAPACITORS IN SERIES AND PARALLEL

Like resistors or inductors, capacitors may be connected in series, in parallel, or in a series-parallel combination. Unlike resistors or inductors, however, total capacitance in series, in parallel, or in a series-parallel combination is found in a different manner. Simply put, the rules are not the same for the calculation of total capacitance. This difference is explained as follows:

Parallel capacitance is calculated like series resistance, and series capacitance is calculated like parallel resistance.

For example:

When capacitors are connected in *series* (see Figure 8.12), the total capacitance C_T is

$$\text{Series: } \frac{1}{C_T} = \frac{1}{C_1} + \frac{1}{C_2} + \frac{1}{C_3} + \ldots + \frac{1}{C_n} \tag{8.5}$$

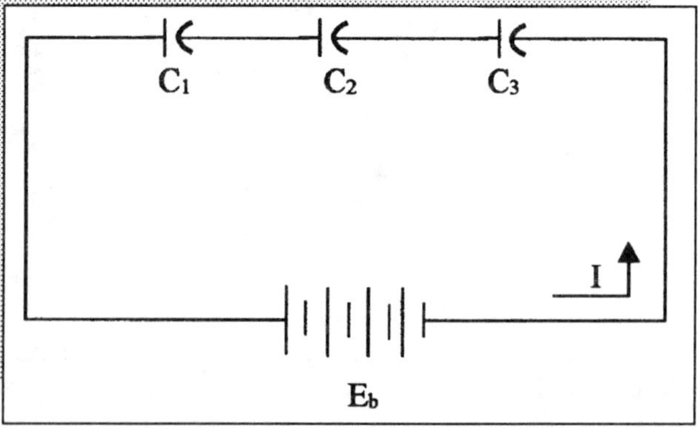

Figure 8.12
Series capacitive circuit.

EXAMPLE 8.2

Problem: Find the total capacitance of a 3-μf, a 5-μf, and a 15-μf capacitor in series.

Solution: Write Equation (8.5) for three capacitors in series.

$$\frac{1}{C_T} = \frac{1}{C_1} + \frac{1}{C_2} + \frac{1}{C_3}$$

$$= \frac{1}{3} + \frac{1}{5} + \frac{1}{15} = \frac{9}{15} = \frac{3}{5} = \frac{5}{3} = 1.7\ \mu f$$

When capacitors are connected in **parallel** (see Figure 8.13), the total capacitance CT is the sum of the individual capacitances.

$$\text{Parallel:} \quad C_T = C_1 + C_2 + C_3 + \ldots + C_n \qquad (8.6)$$

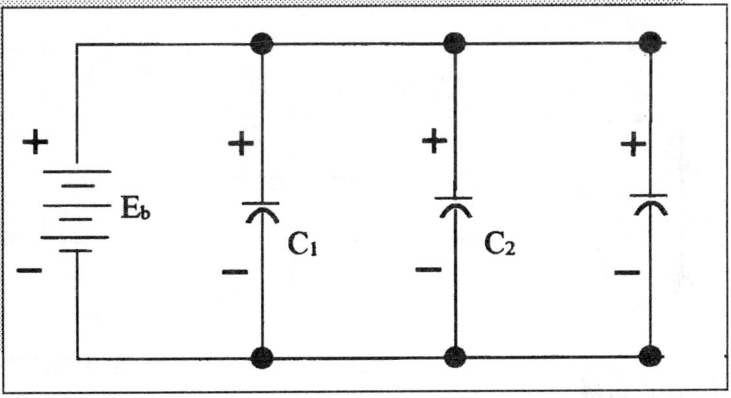

Figure 8.13
Parallel capacitive circuit.

EXAMPLE 8.3

Problem: Determine the total capacitance in a parallel capacitive circuit:

Given:

$$C_1 = 2\ \mu f$$
$$C_2 = 3\ \mu f$$
$$C_3 = 0.25\ \mu f$$

Solution: Write Equation (8.6) for three capacitors in parallel:

$$\begin{aligned} C_T &= C_1 + C_2 + C_3 \\ &= 2 + 3 + 0.25 \\ &= 5.25\ \mu f \end{aligned}$$

If capacitors are connected in a combination of **series and parallel** (see Figure 8.14), the total capacitance is found by applying Equations (8.5) and (8.6) to the individual branches.

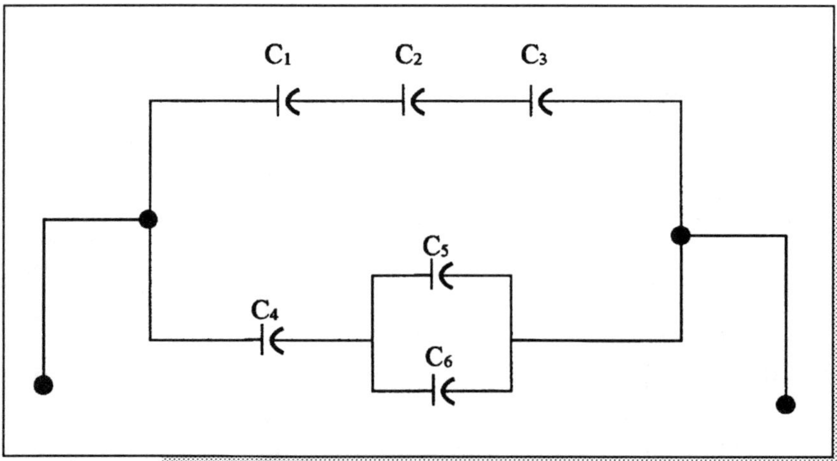

Figure 8.14
Series-parallel capacitance configuration.

8.11 TYPES OF CAPACITORS

Capacitors used for commercial applications are divided into two major groups—fixed and variable—and are named according to their dielectric. Most common are air, mica, paper, and ceramic capacitors, plus the electrolytic type. These types are compared in Table 8.2.

The fixed capacitor has a set value of capacitances that is determined by its construction. The construction of the variable capacitor allows a range of capacitances. Within this range, the desired value of capacitance is obtained by some mechanical means, such as by turning a shaft (as in turning a radio tuner knob, for example) or adjusting a screw to adjust the distance between the plates.

The electrolytic capacitor consists of two metal plates separated by an electrolyte. The electrolyte, either paste or liquid, is in contact with the negative terminal, and this combination forms the negative electrode. The dielectric is a very thin film of oxide deposited on the positive electrode, which is aluminum sheet. Electrolytic capacitors are polarity sensitive (i.e., they must be connected in a circuit according to their polarity markings) and are used where a large amount of capacitance is required.

TABLE 8.2. Comparison of Capacitor Types.

Dielectric	Construction	Capacitance Range
Air	Meshed plates	10–400 pf
Mica	Stacked plates	10–5000 pf
Paper	Rolled foil	0.001–1 μf
Ceramic	Tubular	0.5–1600 pf
	Disk	0.002–0.1 μf
Electrolytic	Aluminum	5–1000 μf
	Tantalum	0.01–300 μf

Self-Test

8.1 Describe a basic capacitor.

8.2 One microfarad is what fraction of a farad?

8.3 Name the three factors that affect capacitance.

8.4 Define the *RC* time constant.

8.5 What is the total capacitance of two 60-microfarad capacitors connected in parallel?

8.6 Fixed capacitors are classified according to the type of material used for the ____________________.

8.7 When using electrolytic capacitors in a circuit, what is an important point that must be remembered?

8.8 A capacitor is used to store up an electric ________________.

8.9 The capacitance of a capacitor is a measure of its ability to store up ________________.

8.10 The capacitance ____________________ if the plate area increases or the dielectric thickness ________________.

Inductive and Capacitive Reactance

TOPICS

Inductive Reactance
Capacitive Reactance
Phase Relationship of R, L, and C Circuits
Impedance
Power in Reactive Circuits

9.1 INTRODUCTION

In Chapters 7 and 8, we learned that the inductance of a circuit acts to oppose any change of current flow in that circuit and that capacitance acts to oppose any change of voltage. In d-c circuits, these **reactions** are not important, because they are momentary and occur only when a circuit is first closed or opened. In a-c circuits, these effects become very important because the direction of current flow is reversed many times each second; and the opposition presented by inductance and capacitance is, for practical purposes, constant.

In purely resistive circuits, either d-c or a-c, the term for opposition to current flow is resistance. When the effects of capacitance or inductance are present, as they often are in a-c circuits, the opposition to current flow is called *reactance.* The total opposition to current flow in circuits that have both resistance and reactance is called *impedance.*

In this chapter, we cover the calculation of inductive and capacitive reactance and impedance; the phase relationships of resistance, inductive, and capacitive circuits; and power in reactive circuits.

Key Terms Used in This Chapter

REACTANCE	The imaginary part of the impedance of an a-c circuit.
IMPEDANCE	The total opposition to current flow in an a-c circuit.
TRUE POWER	Average power actually consumed by an a-c circuit.
APPARENT POWER	The product of the rms volts and rms amps.
POWER FACTOR	The ratio of true power to apparent power.

9.2 INDUCTIVE REACTANCE

In order to gain understanding of the reactance of a typical coil, we need to review exactly what occurs when an a-c voltage is impressed across the coil.

1. The a-c voltage produces an alternating current.
2. When a current flows in a wire, lines of force are produced around the wire.
3. Large currents produce many lines of force; small currents produce only a few lines of force.
4. As the current changes, the number of lines of force will change. The field of force will seem to expand and contract as the current increases and decreases as shown in Figure 9.1.
5. As the field expands and contracts, the lines of force must cut across the wires that form the turns of the coil.
6. These cuttings induce an emf in the coil.
7. This emf acts in the direction so as to oppose the original voltage and is called a *counter* (or back) *emf.*
8. The effect of this counter emf is to reduce the original voltage

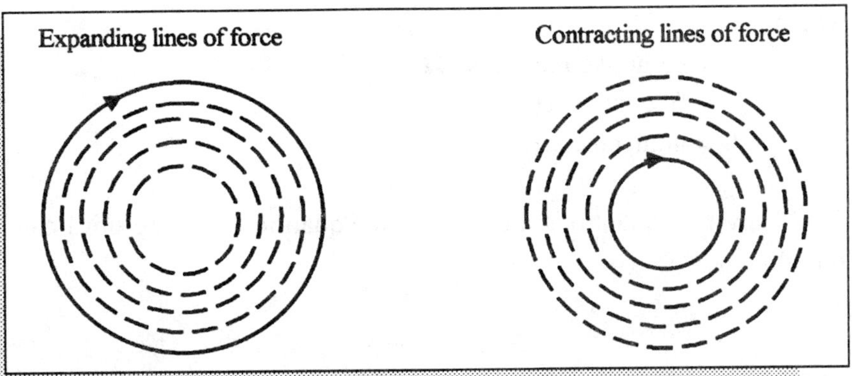

Figure 9.1
An a-c current producing a moving (expanding and collapsing) field. In a coil, this moving field cuts the wires of the coil.

impressed on the coil. The net effect will be to reduce the current below that which would flow if there were no cuttings or counter emf.

9. In this sense, the counter emf is acting as a resistance in reducing the current.
10. Although it would be more convenient to consider the current-reducing effect of a counter emf as a number of ohms of effective resistance, we don't do this. Instead, because a counter emf is not actually a resistance but merely **acts** as a resistance, we use the term *reactance* to describe this effect.

IMPORTANT ***Important Point:*** *The* **reactance** *of a coil is the number of ohms of resistance that the coil seems to offer as a result of a counter emf induced in it. Its symbol is* X *to differentiate it from the d-c resistance* R.

The inductive reactance of a coil depends primarily on (1) the coil's inductance and (2) the frequency of the current flowing through the coil. The value of the reactance of a coil is therefore proportional to its inductance and the frequency of the a-c circuit in which it is used.

The formula for inductive reactance is

$$X_L = 2\pi fL \qquad (9.1)$$

Because $2\pi = 2(3.14) = 6.28$, Equation (9.1) becomes

$$X_L = 6.28fL$$

where

X_L = inductive reactance, Ω
f = frequency, Hz
L = inductance, H

If any two quantities are known in Equation (9.1), the third can be found.

$$L = \frac{X_L}{6.28f} \tag{9.2}$$

$$f = \frac{X_L}{6.28L} \tag{9.3}$$

EXAMPLE 9.1

Problem: The frequency of a circuit is 60 Hz, and the inductance is 20 mh. What is X_L?

Solution:

$$X_L = 2\pi fL$$
$$= 6.28 \times 60 \times 0.02$$
$$= 7.5\ \Omega$$

EXAMPLE 9.2

Problem: A 30-mh coil is in a circuit operating at a frequency of 1400 kHz. Find its inductive reactance.

Solution:

Given: $L = 30$ mh

$f = 1400$ kHz

Find: $X_L = ?$

Step 1: Change units of measurement.

$$30 \text{ mh} = 30 \times 10^{-3} \text{ h}$$

$$1400 \text{ kHz} = 1400 \times 10^{3}$$

Step 2: Find the inductive reactance.

$$X_L = 6.28fL$$

$$X_L = 6.28 \times 1400 \times 10^{3} \times 30 \times 10^{-3}$$

$$X_L = 264{,}000\ \Omega$$

EXAMPLE 9.3

Problem:

Given: $L = 400$ µh

$f = 1500$ Hz

Find: $X_L = ?$

Solution:

$$X_L = 2\pi fL$$

$$= 6.28 \times 1500 \times 0.0004$$

$$= 3.77\ \Omega$$

IMPORTANT

Key Point: *If frequency or inductance varies, inductive reactance must also vary. A coil's inductance does not vary appreciably after the coil is manufactured, unless it is designed as a variable inductor. Thus, frequency is generally the only variable factor affecting the inductive reactance of a coil. The coil's inductive reactance will vary directly with the applied frequency.*

9.3 CAPACITIVE REACTANCE

Previously, we learned that as a capacitor is charged, electrons are drawn from one plate and deposited on the other. As more and more electrons accumulate on the second plate, they begin to act as an opposing voltage that attempts to stop the flow of electrons just as a resistor would do. This opposing effect is called the *reactance* of the capacitor and is measured in ohms. The basic symbol for reactance is X, and the subscript defines the type of reactance. In the symbol for inductive reactance, X_L, the subscript L refers to inductance. Following the same pattern, the symbol for capacitive reactance is X_C.

IMPORTANT

Key Point: Capacitive reactance, *X_C, is the opposition to the flow of a-c current due to the capacitance in the circuit.*

The factors affecting capacitive reactance, X_C, are the size of the capacitor and frequency.

The larger the capacitor, the greater the number of electrons that may be accumulated on its plates. However, because the plate area is large, the electrons do not accumulate in one spot but spread out over the entire area of the plate and do not impede the flow of new electrons onto the plate. Therefore, a large capacitor offers a small reactance. If the capacitance were small, as in a capacitor with a small plate area, the electrons could not spread out and would attempt to stop the flow of electrons coming to the plate. Therefore, a small capacitor offers a large reactance. The reactance is therefore **inversely** proportional to the capacitance.

If an a-c voltage is impressed across the capacitor, electrons are accumulated first on one plate and then on the other. If the frequency of the changes in polarity is low, the time available to accumulate electrons will be large. This means that a large number of electrons will be able to accumulate, which will result in a large opposing effect, or a large reactance. If the frequency is high, the time available to accumulate electrons will be small. This means that there will be only a few electrons on the plates, which will result in a small opposing effect, or a small reactance. The reactance is, therefore, **inversely** proportional to the frequency.

The formula for capacitive reactance is

$$X_C = \frac{1}{2\pi fC} \tag{9.4}$$

with C measured in farads.

EXAMPLE 9.4

Problem: What is the capacitive reactance of a circuit operating at a frequency of 60 Hz, if the total capacitance is 130 µf?

Solution:

$$X_C = \frac{1}{2\pi fC}$$

$$= \frac{1}{6.28 \times 60 \times 0.00013}$$

$$= 20.4\ \Omega$$

9.4 PHASE RELATIONSHIP R, L, AND C CIRCUITS

Unlike a purely resistive circuit (where current rises and falls with the voltage; that is, it neither leads nor lags and current and voltage are in phase), current and voltage are not in phase in inductive and capacitive circuits. This is the case, of course, because occurrences are not quite instantaneous in circuits that have either inductive or capacitive components.

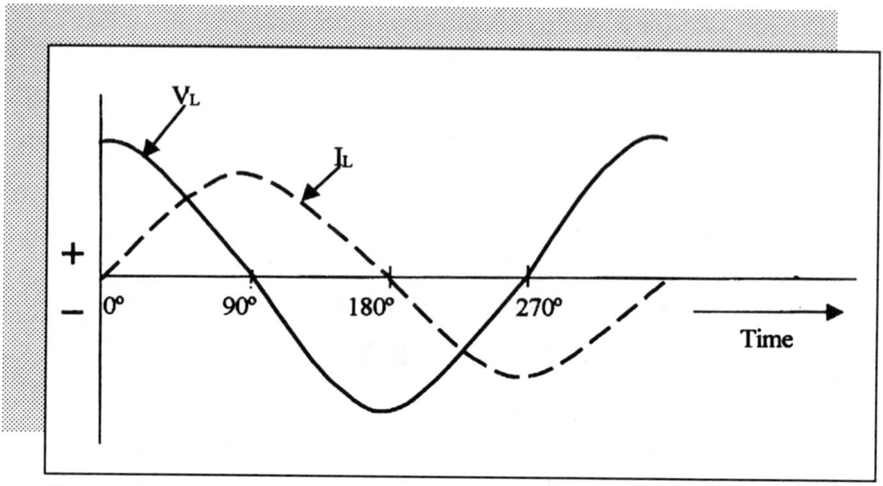

Figure 9.2
Inductive circuit—voltage leads current by 90°.

In the case of an inductor, voltage is first applied to the circuit, then the magnetic field begins to expand, and self-induction causes a counter current to flow in the circuit, opposing the original circuit current. Voltage **leads** current by 90° (see Figure 9.2).

When a circuit includes a capacitor, a charge current begins to flow, and then a difference in potential appears between the plates of the capacitor. Current **leads** voltage by 90° (see Figure 9.3).

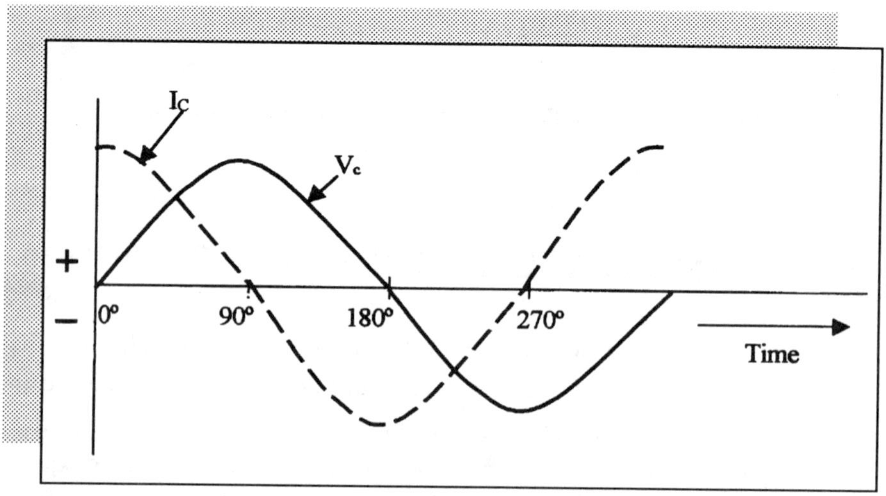

Figure 9.3
Capacitive circuit—current leads voltage by 90°.

Key Point: *In an inductive circuit, voltage leads current by 90°; and in a capacitive circuit, current leads voltage by 90°.*

IMPORTANT

9.5 IMPEDANCE

Impedance is the total opposition to the flow of alternating current in a circuit that contains resistance and reactance. In the case of pure inductance, inductive reactance, X_L, is the total opposition to the flow of current through it. In the case of pure resistance, R represents the total opposition. The combined opposition of R and X_L in series or in parallel to current flow is called *impedance*. The symbol for impedance is Z.

The impedance of resistance in series with inductance is

$$Z = \sqrt{R^2 + X_L^2} \tag{9.5}$$

where

Z = impedance, Ω

R = resistance, Ω

X_L = inductive reactance, Ω

The impedance of resistance in series with capacitance is

$$Z = \sqrt{R^2 + X_C^2} \tag{9.6}$$

where

Z = impedance, Ω

R = resistance, Ω

X_C = inductive capacitance, Ω

When the impedance of a circuit includes R, X_L, and X_C, both resistance and net reactance must be taken into account. The equation for impedance, including both X_L and X_C is

$$Z = \sqrt{R^2 + (X_L - X_C)^2} \tag{9.7}$$

9.6 POWER IN REACTIVE CIRCUITS

The power in a d-c circuit is equal to the product of volts and amps, but in an a-c circuit this is true only when the load is resistive and has no reactance.

In a circuit possessing inductance only, the true power is zero. The current lags the applied voltage by 90°. The true power in a capacitive circuit is also zero. The *true power* is the average power actually consumed by the circuit, the average being taken over one complete cycle of alternating current. The *apparent power* is the product of the rms volts and rms amps.

The ratio of true power to apparent power in an a-c circuit is called the *power factor.* It may be expressed as a percent or as a decimal.

Self-Test

9.1 The reactance of a capacitor is the opposition offered by the capacitor to the passage of an __________ current.

9.2 The reactance decreases as the frequency or capacitance __________.

9.3 What are the factors affecting inductive reactance?

9.4 What are the factors affecting capacitive reactance?

9.5 A circuit whose frequency is 500 Hz has an inductance of 30 mh. $X_L = ?$

9.6 A circuit whose frequency is 200 Hz has a capacitance of 120 μf. $X_C = ?$

9.7 What is the phase relationship between voltage and current in an inductive circuit?

9.8 What is the phase relationship between voltage and current in a capacitive circuit?

9.9 What is the term for all opposition to current flow?

9.10 Define true power:

A-C Circuit Theory

TOPICS

Series RLC Circuits
Parallel RLC Circuits
Power in A-C Circuits

10.1 INTRODUCTION

Chapters 8 and 9 explained how a combination of inductance and resistance and then capacitance and resistance behave in an a-c circuit. We saw how the RL and RC combination affects the current, voltages, power, and power factor of a circuit. We considered these fundamental properties as isolated phenomena. The following phase relationships were seen to be true:

1. The voltage drop across a resistor is in **phase** with the current through it.
2. The voltage drop across an inductor **leads** the current through it by 90°.
3. The voltage drop across a capacitor **lags** the current through it by 90°.
4. The voltage drops across inductors and capacitors are **180° out of phase.**

Solving a-c problems is complicated by the fact that current varies with time as the a-c output of an alternator goes through a complete cycle. This is the case because the various voltage drops in the circuit vary in-phase—they are not at their maximum or minimum values at the same time.

A-c circuits frequently include all three circuit elements: resistance, inductance, and capacitance. In this chapter, all three of these fundamental circuit parameters are combined, and their effect on circuit values is studied.

10.2 SERIES RLC CIRCUITS

Figure 10.1 shows both the sine waveforms and the vectors for purely resistive, inductive, and capacitive circuits. Only the vectors show the

Key Terms Used in This Chapter

RLC CIRCUIT ▶ A circuit containing resistance, inductive reactance, and capacitive reactance.

RESONANCE ▶ When $X_L = X_C$ in a series or parallel RLC circuit.

direction, because the magnitudes are dependent on the values chosen for a given circuit. [**Note:** We are only interested in the **effective** (root-mean square, rms) values]. If the individual resistances and reactances are known, Ohm's Law may be applied to find the voltage drops. For example, we know that $E_R = I \times R$ and $E_C = I \times X_L$. Then, according to Ohm's Law, $E_L = I \times X_L$.

In an a-c circuit, current varies with time; accordingly, the voltage drops across the various elements also vary with time. However, the same variation is not always present in each **at the same time** (except in purely resistive circuits) because current and voltage are not in phase.

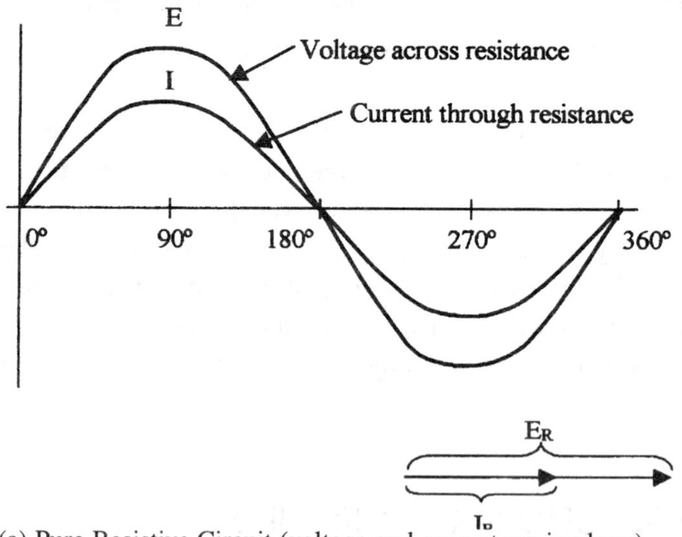

(a) Pure Resistive Circuit (voltage and current are in phase)

Figure 10.1
Sine waveforms and vectorial representation of R, L, and C circuits.

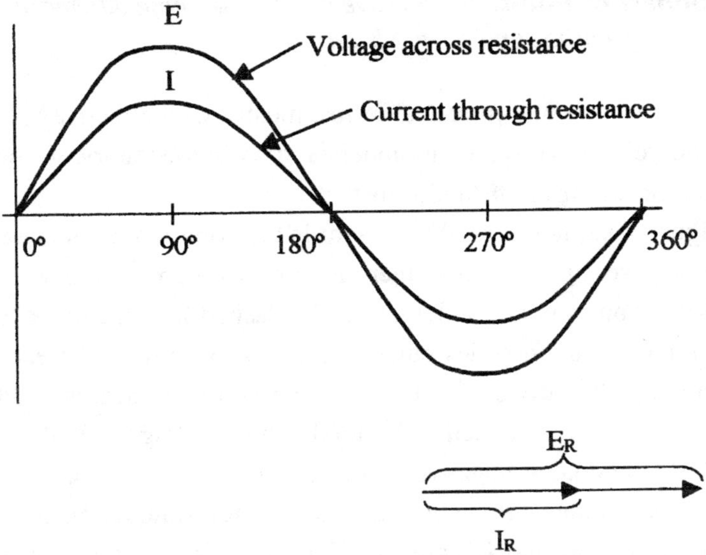

(b) Pure Inductive Circuit (voltage leads current by 90°).

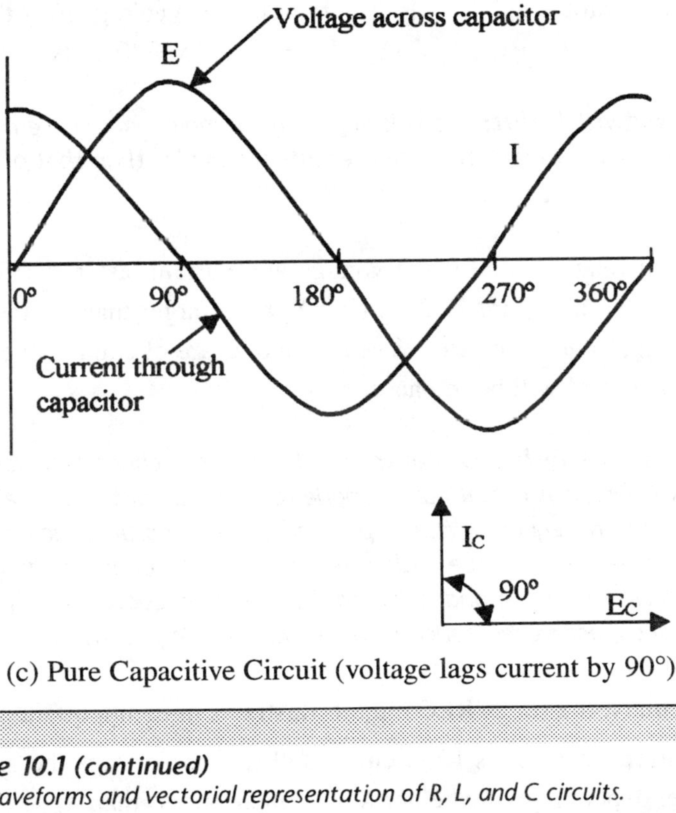

(c) Pure Capacitive Circuit (voltage lags current by 90°).

Figure 10.1 (continued)
Sine waveforms and vectorial representation of R, L, and C circuits.

Important Point: *In a resistive circuit, the phase difference between voltage and current is zero.*

IMPORTANT

We are concerned, in practical terms, mostly with effective values of current and voltage. However, to understand basic a-c theory, we need to know what occurs from instant to instant.

In Figure 10.2, note first that current is the common reference for all three element voltages, because there is only one current in a series circuit, and it is common to all elements. The dashed line in Figure 10.2(a) represents the common series current. The voltage vector for each element, showing its individual relation to the common current, is drawn above each respective element. The total source voltage E is the vector sum of the individual voltages of I_R, IX_L, and IX_C.

The three element voltages are arranged for summation in Figure 10.2(b). Because IX_L and IX_C are each 90° away from I, they are, therefore, 180° from each other. Vectors in direct opposition (180° out of phase) may be subtracted directly. The total reactive voltage E_X is the difference of IX_L and IX_C. Or, $E_X = IX_L - IX_C = 45 - 15 = 30$ volts.

Important Point: *The voltage across a single reactive element in a series circuit can have a greater effective value than that of the applied voltage.*

IMPORTANT

The final relationship of line voltage and current, as seen from the source, is shown in Figure 10.2(c). Had X_C been larger than X_L, the voltage would lag, rather than lead. When X_C and X_L are of equal values, line voltage and current will be in phase.

Important Point: *One of the most important characteristics of an RLC circuit is that it can be made to respond most effectively to a single given frequency. When operated in this condition, the circuit is said to be in* resonance *with or* resonant *to the operating frequency. A circuit is at resonance when the inductive reactance X_L is equal to the capacitive reactance X_C. At resonance, Z equals the resistance R.*

IMPORTANT

In summary, the series RLC circuit illustrates three important points:

1. The current in a series RLC circuit either leads or lags the applied voltage, depending on whether X_C is greater or less than X_L.

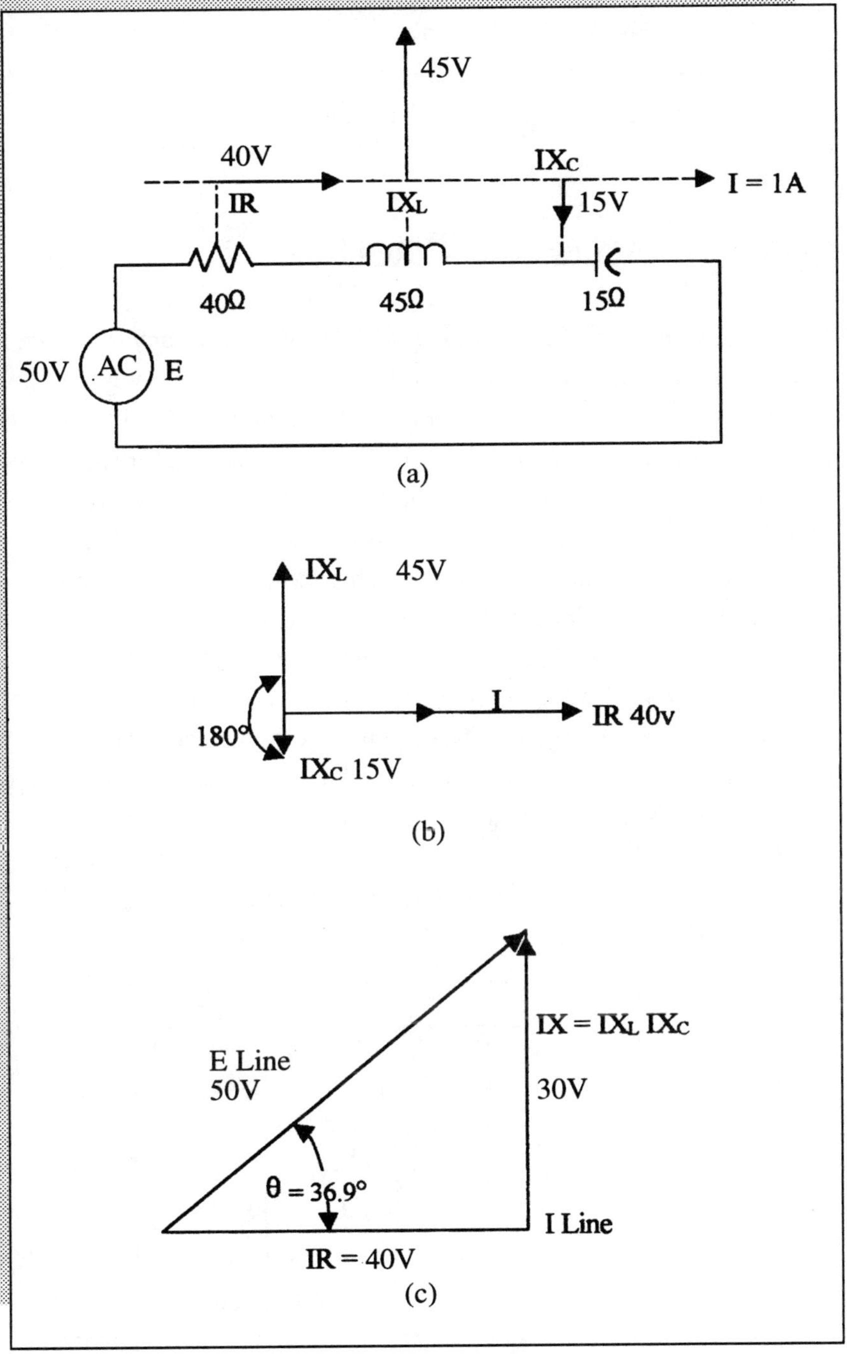

Figure 10.2
Resistance, inductance, and capacitance connected in a series.

2. A capacitive voltage drop in a series circuit always subtracts directly from an inductive voltage drop.
3. The voltage across a single reactive element in a series circuit can have a greater effective value than that of the applied voltage.

10.3 PARALLEL RLC CIRCUITS

The **true power** of a circuit is $P = EI \cos \theta$; and for any given amount of power to be transmitted, the current, I, varies inversely with the power factor, $\cos \theta$. Thus, the addition of capacitance in parallel with inductance will, under the proper conditions, improve the power factor (make nearer to unity power factor) of the circuit and make possible the transmission of electric power with reduced line loss and improved voltage regulation.

Figure 10.3(a) shows a three-branch parallel a-c circuit with a resistance in one branch, inductance in the second branch, and capacitance in the third branch. The voltage is the same across each parallel branch, so $V_T = V_R = V_L = V_C$. The applied voltage V_T is used as the reference line to measure phase angle θ. The total current I_T is the vector sum of I_R, I_L, and I_C. The current in the resistance I_R is in phase with the applied voltage V_T [(see Figure 10.3(b)]. The current in the capacitor I_C leads the voltage V_T by 90°. I_L and I_C are exactly 180° out of phase and, thus, are acting in opposite directions [see Figure 10.3(b)]. When $I_C > I_C$, I_T lags V_T [see Figure 10.3(c)] so the parallel RLC circuit is considered inductive.

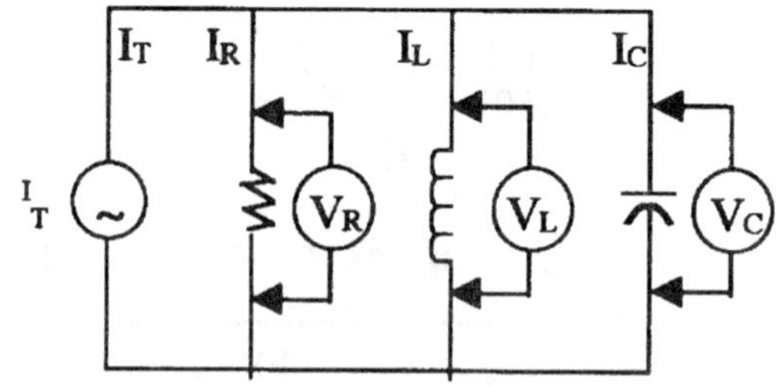

(a) Parallel RLC circuit diagram

Figure 10.3
R, X_L, and X_C in parallel.

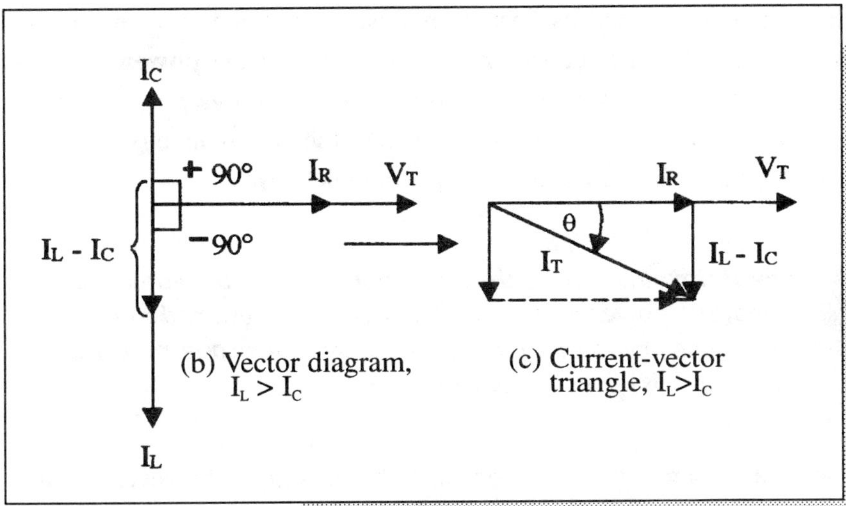

Figure 10.3 continued
R, X_L, and X_C in parallel.

10.4 POWER IN A-C CIRCUITS

In circuits that have only resistance, but no reactance, the amount of power absorbed in the circuit is easily calculated by $P = I^2R$. However, in dealing with circuits that include inductance and capacitance (or both), which is often the case in a-c electricity, the calculation of power is a more complicated process.

In Chapter 3, we explained that power is a measure of the rate at which work is done. The work of a resistor is to limit current flow to the correct, safe level. In accomplishing this, the resistor dissipates heat, and we say that power is consumed or absorbed by the resistor.

Inductors and capacitors also oppose current flow, but they do so by producing current that opposes the line current. In either inductive or capacitive circuits, instantaneous values of power may be very large, but the power actually absorbed is essentially zero, because only resistance dissipates heat (absorbs power). Both inductance and capacitance return the power to the source.

Any component that has resistance, such as a resistor or the wiring of an inductor, consumes power. Such power is not returned to the source, because it is dissipated as heat. Previously, we stated that power consumed in the circuit is called **true power,** or **average power.** The two terms are interchangeable, but we use the term *average power,* because the overall value is more meaningful than the instantaneous values of power appearing in the circuit during a complete cycle.

IMPORTANT ***Key Point:*** *In terms of the dissipation of power as heat in a circuit,* ***apparent power*** *includes both power that is returned to the source and power that is dissipated as heat.* ***Average power*** *is power that is dissipated as heat.*

Not all apparent power is consumed by the circuit; however, because the alternator does deliver the power, it must be considered in the design. The average power consumption may be small, but instantaneous values of voltage and current are often very large. Apparent power is an important design consideration, especially in assessing the amount of insulation necessary.

In an a-c circuit that includes both reactance and resistance, some power is consumed by the load, and some is returned to the source. How much of each depends on the phase angle, because current normally leads or lags voltage by some angle.

IMPORTANT ***Note:*** *Recall that in a purely reactive circuit, current and voltage are 90° out of phase.*

In an RLC circuit, the ratio of R/Z is the cosine of the phase angle θ. Therefore, it is easy to calculate average power in an RLC circuit:

$$P = EI \cos \theta \qquad (10.1)$$

where

E = effective value of the voltage across the circuit
I = effective value of current in the circuit
θ = phase angle between voltage and current
P = average power absorbed by the circuit

IMPORTANT **Note:** *Recall that the equation for average power in a purely resistive circuit is* P = EI. *In a resistive circuit,* P = EI, *because the cos θ is 1 and need not be considered. In most cases, the phase angle will be neither 90 nor zero, but somewhere between those extremes.*

EXAMPLE 10.1

Problem: An RLC circuit has a source voltage of 500 volts, line current is 2 amps, and current leads voltage by 60°. What is the average power?

Solution:

Average Power = 500 V x 2 A x 0.5

(Note: cos of 60° = 0.5)

= 500 watts

EXAMPLE 10.2

Problem: An RLC circuit has a source voltage of 300 volts, line current is 2 amps, and current lags voltage by 31.8°. What is the average power?

Solution:

Average Power = 300 V x 2 A x 0.8499

= 509.9 watts

EXAMPLE 10.3

Problem:

Given:

$E = 100$ volts

$I = 4$ amps

$\theta = 58.4$

What is the average power?

Solution:

Average Power = 100 V x 4 A x 0.5240

= 209.6 watts

Self-Test

10.1 What kind of power is associated with purely resistive circuits?

10.2 What kind of power is associated with reactive circuits (excluding resistance)?

10.3 In a purely reactive circuit, current and voltage are _________ degrees out of phase.

10.4 What is cos θ for a purely reactive circuit?

10.5 Because cos θ is zero, what is the average power in a purely reactive circuit?

10.6 What is the cos θ in a purely resistive circuit?

10.7 What two things must be known before vectors can be used to add voltages or currents in an RLC circuit?

10.8 Under what conditions is an RLC circuit at resonance?

10.9 Define power factor:

10.10 At resonance, Z_t is the same as __________.

Practical Electrical Applications

TOPICS

Electrical Power Generation
Motors
Transformers
Power Distribution System Protection

11.1 INTRODUCTION

In Chapter 1 (Section 1.1), we pointed out that water and/or wastewater operators normally have little difficulty recognizing electrical equipment within their plant sites. This is the case, of course, because there are few places within the plant site (that is, in the majority of plant sites) that an operator can go where electricity is not performing some important function. Simply stated, whether the important function be powering lighting, heating, air conditioning, pump motors, mechanized bar screens, control systems, communications equipment, or computerized systems, it would be difficult for the modern operator to imagine plant operations without the use of electrical power.

To this point in the manual, we have concentrated on the fundamentals of electricity and electric circuits. This was our goal. Along with satisfying our goal, we understand that having a basic knowledge of electrical theory is a great accomplishment. However, a knowledge of basic theory (of any type) that is not put to practical use is analogous to understanding the operation of an internal combustion engine without ever having the opportunity to work on one.

In short, for the water/wastewater operator, having an understanding of basic electrical fundamentals assists in the successful passing of various certification examinations—no doubt an important career-enhancing achievement. Certification is, however, just one critical element (one important step) required of operators. Operators must also be qualified to

Key Terms Used in This Chapter

Generator	A device that produces voltage output.
Motor	A device that produces mechanical motion from electric energy.
Transformer	A device that has two or more coil windings used to step up or step down a-c voltage.
Mutual Induction	The ability of one coil to induce voltage in another coil.
Fuse	A metal link that melts from excessive current and opens circuit.
Circuit Breaker	A protective device that opens when excessive current flows in circuit.

operate the plant and its associated machinery—much of which is electrical equipment. To this end, we have incorporated (along with the required theory) pertinent information on electrical applications most important to the operator in his or her daily task of operating plant electrical equipment as it was intended to be operated—as it should be operated; that is, operated "with understanding."

11.2 ELECTRICAL POWER GENERATION

Most water/wastewater treatment plants do not generate their own plant general service electrical power. Instead, and as with most other industrial users, treatment plants typically purchase electrical power from a local electrical utility company, and any on-site generation is provided only for standby power. Additionally, in smaller communities, it is not unusual for a pumping station to be located in residential areas where an associated emergency generator has been installed as an integral part of the station. Portable generators have also seen increased use in recent years. Generators used for standby power will normally be the gasoline, diesel, or gas-turbine types, because these can be started immediately.

IMPORTANT ***Note:*** *Most electrical utility companies operate with a reliability in excess of 90+%. However, even a momentary loss of electrical power to plant operations cannot, in many instances, be tolerated. Therefore, standby electrical generators are important in maintaining plant operations. One of the main reasons why standby power is installed in plants is to provide emergency lighting for safe egress in the event of a failure of the utility supply. Exit lights, stairway lighting, and a portion of the corridor lighting systems are typically connected to the standby generator.*

Fire and safety equipment often is connected to the standby power system to assist in emergency operations. Fire pumps, communications systems, fire detection and alarm systems, and security systems remain operational in the event of a disruption of the normal electrical supply.

Finally, critical pieces of mechanical equipment, such as pumps, are normally connected to the standby power system to protect the facility from damage if the equipment were out of service during an electrical outage.

Generators can be designed to supply small amounts of power, or they can be designed to supply many thousands of kilowatts of power. Also, generators may be designed to supply either direct current or alternating current.

11.2.1 D-C GENERATORS

A *d-c generator* is a rotating machine that converts mechanical energy into electrical energy. This conversion is accomplished by rotating an armature, which carries conductors, in a magnetic field, thus inducing an emf in the conductors. As stated previously, in order for an emf to be induced in the conductors, a relative motion must always exist between the conductors and the magnetic field in such a manner that conductors cut through the field. In most d-c generators, the armature is the rotating member, and the field is the stationary member. A mechanical force is applied to the shaft of the rotating member to cause the relative motion. Thus, when mechanical energy is put into the machine in the form of a mechanical force or twist on the shaft, causing the shaft to turn at a certain speed, electrical energy in the form of voltage and current is delivered to the external load circuit.

IMPORTANT ***Important Point:*** *Mechanical power must be applied to the shaft constantly so long as the generator is supplying electrical energy to the external load circuit.*

To gain a basic understanding of the operation of a d-c generator, consider the following explanation.

A simple d-c generator consists of an armature coil with a single turn of wire [see Figure 11.1(a) and (b)]. [**Note:** The armature coils used in large d-c machines are usually wound in their final shape before being put on the armature. The sides of the preformed coil are placed in the slots of the laminated armature core.] This armature coil cuts across the magnetic field to produce voltage. If a complete path is present, current will move through the circuit in the direction shown by the arrows [see Figure 11.1(a)]. In this position of the coil, commutator segment 1 is in contact with brush 1, while commutator segment 2 is in contact with brush 2. As the armature rotates a half turn in a clockwise direction, the contacts between the commutator segments and the brushes are reversed [see Figure 11.1(b)]. At this moment, segment 1 is in contact with brush 2 and segment 2 is in contact with brush 1. Because of this commutator action, that side of the armature coil that is in contact with either of the brushes is always cutting across the magnetic field in the same direction. Thus, brushes 1 and 2 have constant polarity, and a **pulsating d-c current** is delivered to the external load circuit.

IMPORTANT ***Note:*** *In d-c generators, voltage induced in individual conductors is a-c. It is converted to d-c (rectified) by the commutator that rotates in contact with carbon brushes so that current generated is in one direction, or direct current.*

There are several different types of d-c generators. They take their names from the type of field excitation used (i.e., they are classified according to the manner in which the field windings are connected to the armature circuit). For example, when the generator's field is excited (or

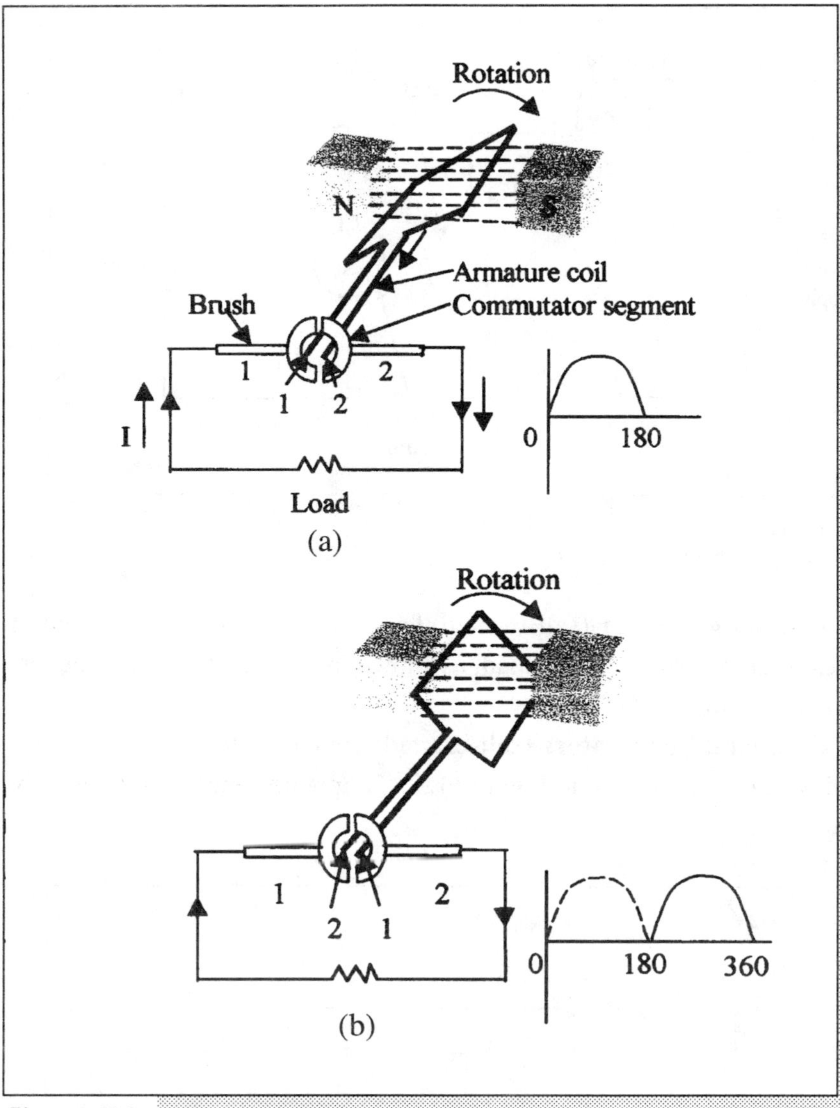

Figure 11.1
Basic operation of a d-c generator.

supplied) from a separate d-c source (such as a battery) other than its own armature, it is called a **separately excited d-c generator** (see Figure 11.2).

A **shunt generator** (self-excited) has its field windings connected in series with a rheostat, across the armature in shunt with the load, as shown in Figure 11.3. The shunt generator is widely used in industry.

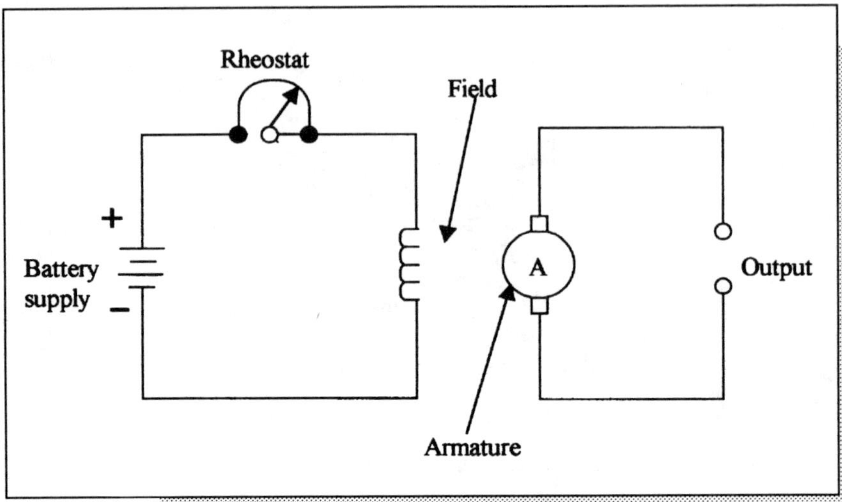

Figure 11.2
Separately excited d-c generator.

A **series generator** (self-excited) has its field windings connected in series with the armature and load, as shown in Figure 11.4. Series generators are seldom used.

Compound generators (self-excited) contain both series and shunt field windings, as shown in Figure 11.5. Compound generators are widely used in industry.

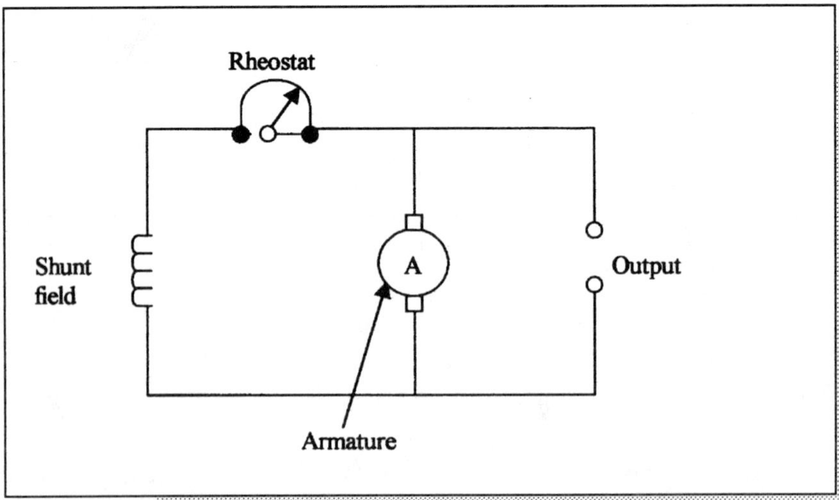

Figure 11.3
A d-c shunt generator.

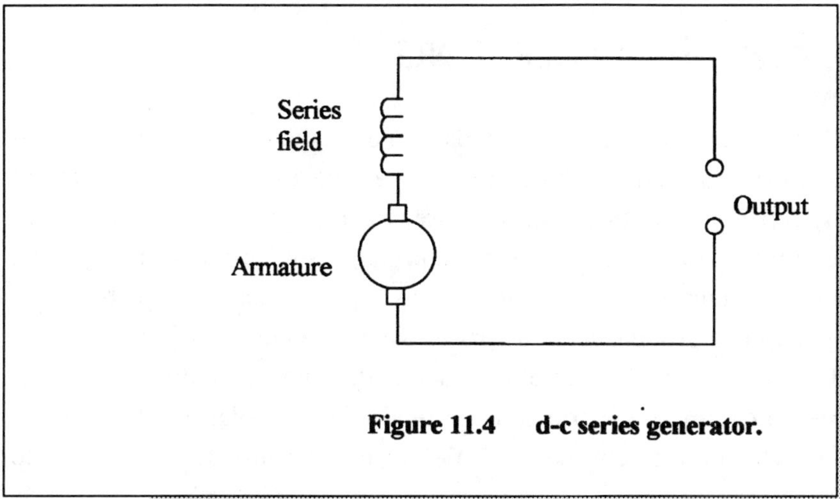

Figure 11.4
A d-c series generator.

Note: *As central generating stations increased in size along with number and power distribution distances, d-c generating systems, because of the high power losses in long d-c transmission lines, were replaced by a-c generating systems to reduce power transmission costs.*

IMPORTANT

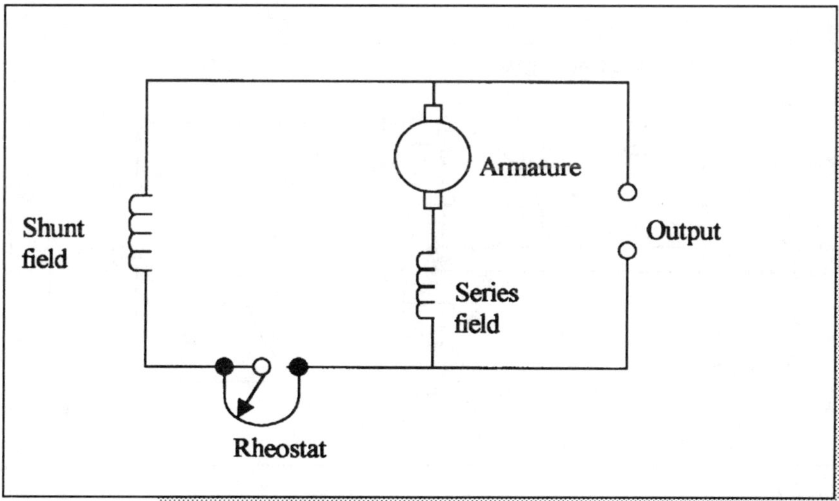

Figure 11.5
A d-c compound generator.

11.2.2 A-C GENERATORS

Most electric power utilized today is generated by *alternating-current generators* (also called **alternators**). They are made in many different sizes, depending on their intended use. Regardless of size, however, all generators operate on the same basic principle—a magnetic field cutting through conductors, or conductors passing through a magnetic field. They are (1) a group of conductors in which the output voltage is generated and (2) a second group of conductors through which direct current is passed to obtain an electromagnetic field of fixed polarity. The conductors in which the electromagnetic field originates are always referred to as the field windings.

In addition to the armature and field, there must also be motion between the two. To provide this, a-c generators are built in two major assemblies, the **stator** and the **rotor.** The rotor rotates inside the stator.

The revolving-field a-c generator (see Figure 11.6) is the most widely used type. In this type of generator, direct current from a separate source is passed through windings on the rotor by means of sliprings and brushes. [**Note:** Sliprings and brushes are adequate for the d-c field supply because the power level in the field is much smaller than in the armature circuit.]

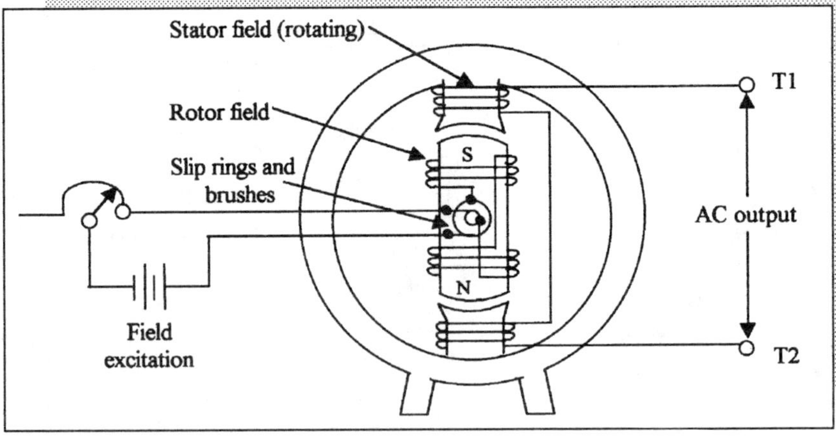

Figure 11.6
Essential parts of a rotating field a-c generator.

This maintains a rotating electromagnetic field of fixed polarity. The rotating magnetic field, following the rotor, extends outward and cuts through the armature windings imbedded in the surrounding stator. As the rotor turns, a-c voltages are induced in the windings because magnetic fields of first one polarity and then the other cut through them. Because the output power is taken from stationary windings, the output may be connected through fixed output terminals T1 and T2 in Figure 11.6. This is advantageous, in that there are no sliding contacts and the whole output circuit is continuously insulated.

IMPORTANT **Important Point:** *In a-c generators, frequency and electromagnetic wave cycles per second depend on how fast the rotor turns and the number of electromagnetic field poles. Voltage generated depends on the rotor speed, number of coils in the armature, and strength of the magnetic field.*

11.3 MOTORS

At least 60% of the electrical power fed to a typical waterworks and/or wastewater treatment plant is consumed by electric motors. One thing is certain: there is an almost endless variety of tasks that electric motors perform in water/wastewater treatment.

An *electric motor* is a machine used to change electrical energy to mechanical energy to do the work. [Note: Recall that a generator does just the opposite; that is, a generator changes mechanical energy to electrical energy.]

Previously, we pointed out that when a current passes through a wire, a magnetic field is produced around the wire. If this magnetic field passes through a stationary magnetic field, the fields either repel or attract, depending on their relative polarity. If both are positive or negative, they repel. If they are opposite polarity, they attract.

Applying this basic information to motor design, an electromagnetic coil, the armature, rotates on a shaft. The armature and shaft assembly are called the rotor. The rotor is assembled between the poles of a permanent magnet, and each end of the rotor coil (armature) is connected to a commutator also mounted on the shaft. A commutator is composed of copper segments insulated from the shaft and from each other by an insulating material. As like poles of the electromagnet in the rotating armature pass the stationary permanent magnet poles, they are repelled, continuing the motion. As the opposite poles near each other, they attract, continuing the motion.

11.3.1 D-C MOTORS

The construction of a d-c motor is essentially the same as that of a d-c generator. However, it is important to remember that the d-c generator converts mechanical energy into the electrical energy back into mechanical energy. A d-c generator may be made to function as a motor by applying a suitable source of d-c voltage across the normal output electrical terminals.

There are various types of d-c motors, depending on the way the field coils are connected. Each has characteristics that are advantageous under given load conditions.

Shunt motors (see Figure 11.7) have the field coils connected in parallel with the armature circuit. This type of motor, with constant potential applied, develops variable torque at an essentially constant speed, even under changing load conditions. Such loads are found in machine-shop equipment, such as lathes, shapes, drills, milling machines, and so forth.

Series motors (see Figure 11.8) have the field coils connected in series with the armature circuit. This type of motor, with constant potential applied, develops variable torque but its speed varies widely under

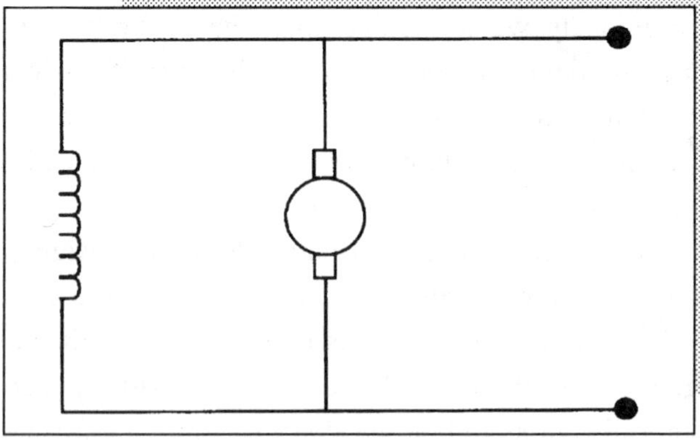

Figure 11.7
A d-c shunt motor.

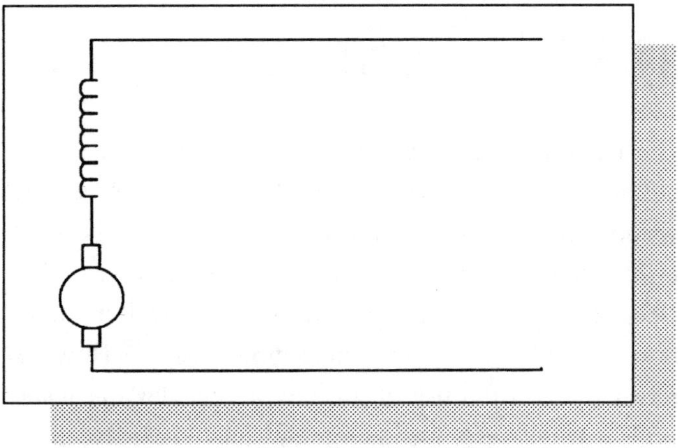

Figure 11.8
A d-c series motor.

changing load conditions. That is, the speed is low under heavy loads, but becomes excessively high under light loads. Series motors are commonly used to drive electric hoists, winches, cranes, and certain types of vehicles (e.g., electric trucks). Also, series motors are used extensively to start internal combustion engines.

Compound motors (see Figure 11.9) have one set of field coils in parallel with the armature circuit and another set of field coils in series with the armature circuit. This type of motor is a compromise between shunt and series motors. It develops an increased starting torque over that of the shunt motor and has less variation in speed than the series motor.

The speed of a d-c motor is variable. It is increased or decreased by a rheostat connected in series with the field or in parallel with the rotor. Interchanging either the rotor or field winding connections reverses direction.

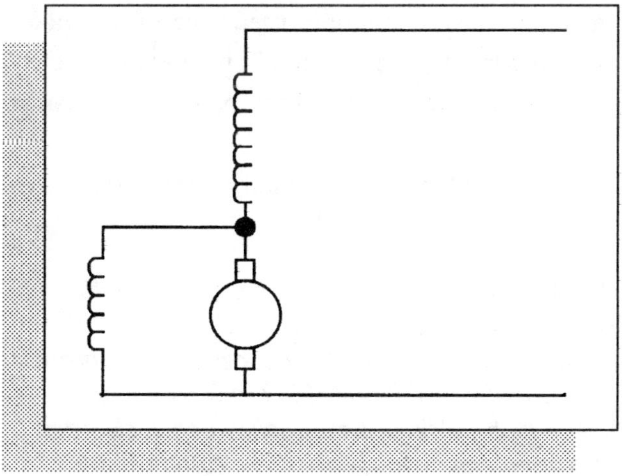

Figure 11.9
A d-c compound motor.

11.3.2 A-C MOTORS

The a-c voltage can be easily transformed from low voltages to high voltages or vice versa, and can be moved over a much greater distance without too much loss in efficiency. Most of the power generating systems today, therefore, produce alternating current. Thus, it logically follows that a great majority of the electrical motors utilized today are designed to operate on alternating current. However, there are other advantages in the use of a-c motors besides the wide availability of a-c power. In general, a-c motors are less expensive than d-c motors. Most types of a-c motors do not employ brushes and commutators. This eliminates many problems of maintenance and wear and, also, eliminates dangerous sparking.

A-c motors are manufactured in many different sizes, shapes, and ratings, for use on an even greater number of jobs. They are designed for use with either polyphase or single-phase power systems.

This chapter cannot possibly cover all aspects of the subject of a-c motors. Consequently, it will deal mainly with the operating principles of the two most common types—the induction and the synchronous motor.

11.3.2.1 INDUCTION MOTORS (POLYPHASE)

The induction motor is the most commonly used type of a-c motor because of its simple, rugged construction and good operating characteristics. It consists of two parts: the **stator** (stationary part) and the **rotor** (rotating part).

The most important type of polyphase induction motor is the three-phase motor.

IMPORTANT

Important Note: *A three-phase (3-θ) system is a combination of three single-phase (1-θ) systems. In a 3-θ balanced system, the power comes from an a-c generator that produces three separate but equal voltages, each of which is out of phase with the other voltages by 120°. Although 1-θ circuits are widely used in electrical systems, most generation and distribution of a-c current is 3-θ.*

The driving torque of both d-c and a-c motors is derived from the reaction of current-carrying conductors in a magnetic field. In the d-c motor, the magnetic field is stationary, and the armature, with its current-carrying conductors, rotates. The current is supplied to the armature through a commutator and brushes.

In *induction motors,* the rotor currents are supplied by electromagnet induction. The stator windings, connected to the a-c supply, contain two or more out-of-time-phase currents, which produce corresponding mmfs. These mmfs establish a rotating magnetic field across the airgap. This magnetic field rotates continuously at constant speed regardless of the load on the motor. The stator winding corresponds to the armature winding of a d-c motor or to the primary winding of a transformer (see Section 11.4). The rotor is not connected electrically to the power supply.

The induction motor derives its name from the fact that mutual induction (or transformer action) takes place between the stator and the rotor under operating conditions. The magnetic revolving field produced by the stator cuts across the rotor conductors, inducing a voltage in the conductors. This induced voltage causes rotor current to flow. Hence, motor torque is developed by the interaction of the rotor current and the magnetic revolving field.

11.3.2.2 SYNCHRONOUS MOTORS

Like induction motors, *synchronous motors* have stator windings that produce a rotating magnetic field. But unlike the induction motor, the synchronous motor requires a separate source of d-c from the field. It also requires special starting components. These include a salient-pole field with starting grid winding. The rotor of the conventional type of synchronous motor is essentially the same as that of the salient-pole a-c generator. The stator windings of induction and synchronous motors are essentially the same.

In operation, the synchronous motor rotor locks into step with the rotating magnetic field and rotates at the same speed. If the rotor is pulled out of step with the rotating stator field, no torque is developed, and the motor stops. Because a synchronous motor develops torque only when running at synchronous speed, it is not self-starting and, hence, needs some device to bring the rotor to synchronous speed. For example, a synchronous motor may be started rotating with a d-c motor on a common shaft. After the motor is brought to synchronous speed, a-c current is applied to the stator windings. The d-c starting motor now acts as a d-c generator, which supplies d-c field excitation for the rotor. The load then can be coupled to the motor.

11.3.2.3 SINGLE-PHASE MOTORS

Single-phase (1-θ) motors are so called because their field windings are connected directly to a single-phase source. These motors are used extensively in fractional horsepower sizes in commercial and domestic applications. The advantages of using single-phase motors in small sizes are that they are less expensive to manufacture than other types, and they eliminate the need for three-phase a-c lines. Single-phase motors are used in fans, refrigerators, portable drills, grinders, and so forth.

A single-phase induction motor with only one stator winding and a cage rotor is like a three-phase induction motor with a cage rotor except that the single-phase motor has no magnetic revolving field at start, and hence, no starting torque. However, if the rotor is brought up to speed by external means, the induced currents in the rotor will cooperate with the stator currents to produce a revolving field, that causes the rotor to continue to run in the direction that it was started.

Several methods are used to provide the single-phase induction motor with starting torque. These methods identify the motor as split-phase, capacitor, shaded-pole, and repulsion-start induction motor.

Another class of single-phase motors is the a-c series (universal) type. Only the more commonly used types of single-phase motors are described. These include the (1) split-phase motor, (2) capacitor motor, (3) shaded-pole motor, (4) repulsion-start motor, and (5) a-c series motor.

11.3.2.3.1 SPLIT-PHASE MOTOR

The *split-phase motor* (see Figure 11.10) has a stator composed of slotted lamination that contains a starting winding and a running winding.

Note: *If two stator windings of unequal impedance are spaced 90 electrical degrees apart but are connected in parallel to a single-phase source, the field produced will appear to rotate. This is the principle of* **phase splitting.**

IMPORTANT

The starting winding has fewer turns and smaller wire than the running winding. Hence, it has higher resistance and less reactance. The main winding occupies the lower half of the slots, and the starting winding occupies the upper half. When the same voltage is applied to both windings, the current in the main winding lags behind the current in the starting winding. The angle *q* between the main and starting windings is enough phase difference to provide a weak rotating magnetic field to produce a starting torque. When the motor reaches a predetermined speed, usually 75% of synchronous speed, a centrifugal switch mounted on the motor shaft opens, thereby disconnecting the starting winding.

Because it has a low starting torque, fractional-horsepower split-phase motors are used in a variety of equipment, such as washers, oil burners, ventilating fans, and woodworking machines. The direction of rotation of the split-phase motor can be reversed by interchanging the starting winding leads.

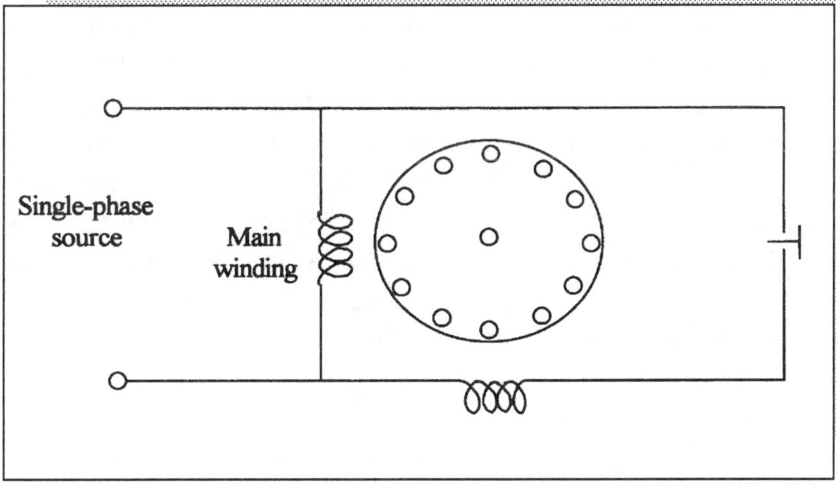

Figure 11.10
Split-phase motor.

11.3.2.3.2 CAPACITOR MOTOR

The *capacitor motor* is a modified form of split-phase motor, having a capacitor in series with the starting winding (see Figure 11.11). The capacitor motor operates with an auxiliary winding and series capacitor permanently connected to the line (see Figure 11.11). The capacitance in series may be of one value for starting and another value for running. As the motor approaches synchronous speed, the centrifugal switch disconnects one section of the capacitor.

If the starting winding is cut out after the motor has increased in speed, the motor is called a **capacitor-start motor.** If the starting winding and capacitor are designed to be left in the circuit continuously, the motor is called a **capacitor-run motor.** Capacitor motors are used to drive grinders, drill presses, refrigerator compressors, and other loads that require relatively high starting torque. The direction of rotation of the capacitor motor may be reversed by interchanging the starting winding leads.

11.3.2.3.3 SHADED-POLE MOTOR

A *shaded-pole motor* employs a salient-pole stator and a cage rotor. The projecting poles on the stator resemble those of d-c machines except that the entire magnetic circuit is laminated and a portion of each pole is split to accommodate a short-circuited coil called a **shading coil** (see Figure 11.12). The coil is usually a single band or strap of copper. The effect of the coil is to produce a small sweeping motion of the field flux from one side of the pole piece to the other as the field pulsates. This slight shift in the magnetic field produces a small starting torque. Thus, shaded-pole motors are self-starting. This motor is generally manufactured in very small sizes, up to 1/20 horsepower, for driving small fans, small appliances, and clocks.

In operation, during that part of the cycle when the main pole flux is increasing, the shading coil is cut by the flux, and the resulting induced emf and current in the shading coil tend to prevent the flux from rising

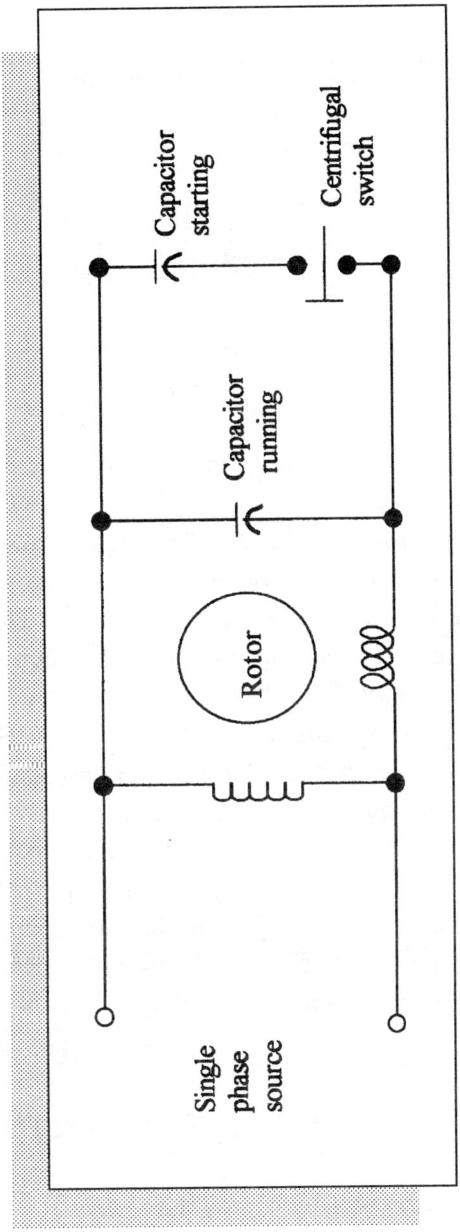

Figure 11.11
Capacitor motor.

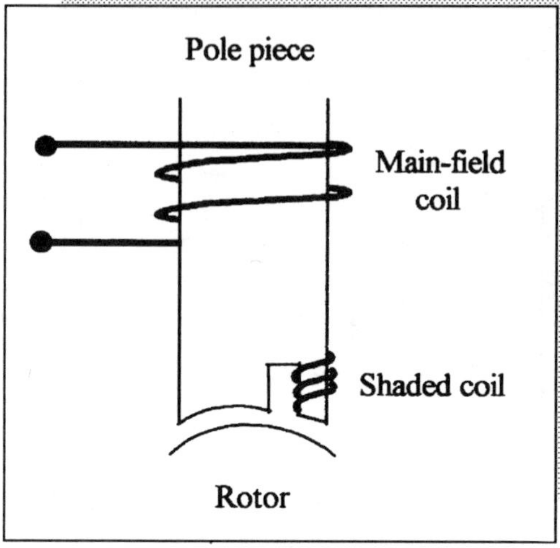

Figure 11.12
Shaded pole.

readily through it. Thus, the greater portion of the flux rises in that portion of the pole that is not in the vicinity of the shading coil. When the flux reaches its maximum value, the rate of change of flux is zero, and the voltage and current in the shading coil are also zero. At this time, the flux is distributed more uniformly over the entire pole face. Then, as the main flux decreases toward zero, the induced voltage and current in the shading coil reverse their polarity, and the resulting mmf tends to prevent the flux from collapsing through the iron in the region of the shading coil. The result is that the main flux first rises in the unshaded portion of the pole and later in the shaded portion. This action is equivalent to a sweeping movement of the field across the pole face in the direction of the shaded pole. The rotor conductors are cut by this moving field, and the force exerted on them causes the rotor to turn in the direction of the sweeping field. The shaded-pole method of starting is used in very small motors, up to about 1/25 hp, for driving small fans, small appliances, and clocks.

11.3.2.3.4 REPULSION-START MOTOR

Like a d-c motor, the *repulsion-start motor* has a form-wound rotor with commutator and brushes. The stator is laminated and contains a distributed single-phase winding. In its simplest form, the stator resembles that of the single-phase motor. In addition, the motor has a centrifugal device that removes the brushes from the commutator and places a short-circuiting ring around the commutator. This action occurs at about 75% of synchronous speed. Thereafter, the motor operates with the characteristics of the single-phase induction motor. This type of motor is made in sizes ranging from 1/2 to 15 hp and is used in applications requiring a high starting torque.

11.3.2.3.5 A-C SERIES MOTOR

The *a-c series motor* will operate on either a-c or d-c circuits. When an ordinary d-c series motor is connected to an a-c supply, the current drawn by the motor is low due to the high series-field impedance. The result is low running torque. To reduce the field reactance to a minimum, a-c series motors are built with as few turns as possible. Armature reaction is overcome by using **compensating windings** (see Figure 11.13) in the pole pieces.

As with d-c series motors, in an a-c series motor, the speed increases to a high value with a decrease in load. The torque is high for high armature currents so that the motor has a good starting torque. A-c series motors operate more efficiently at low frequencies.

Fractional horsepower a-c series motors are called **universal motors.** They do not have compensating windings. They are used extensively to operate fans and portable tools, such as drills, grinders, and saws.

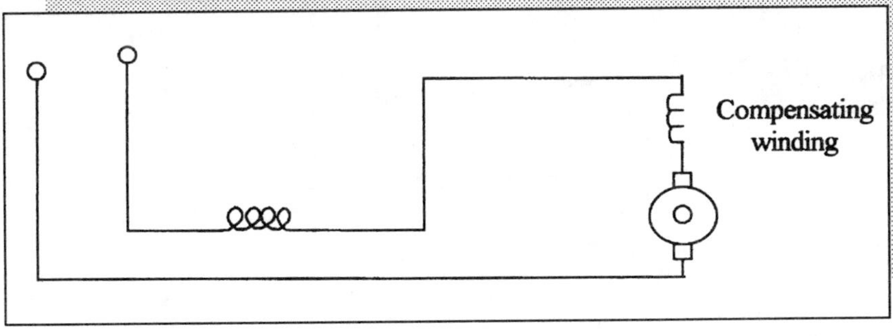

Figure 11.13
An a-c series motor.

11.4 TRANSFORMERS

A *transformer* is an electric control device (with no moving parts) that raises or lowers voltage or current in an electric distribution system. The basic transformer consists of two coils that are electrically insulated from each other and wound upon a common core (see Figure 11.14). Magnetic coupling is used to transfer electric energy from one coil to another. The coil that **receives** energy from an a-c source is called the **primary.** The coil that delivers energy to an a-c load is called the **secondary.** The core of transformers used at low frequencies is generally made of magnetic material, usually laminated sheet steel. Cores of transformers used at higher frequencies are made of powdered iron and ceramics, or nonmagnetic materials. Some coils are simply wound on nonmagnetic hollow forms such as cardboard or plastic so that the core material is actually air.

In operation, an alternating current will flow when an a-c voltage is applied to the primary coil of a transformer. This current produces a field of force that changes as the current changes. The changing magnetic field is carried by the magnetic core to the secondary coil, where it cuts across the turns of that coil. In this way, an a-c voltage in one coil is transferred

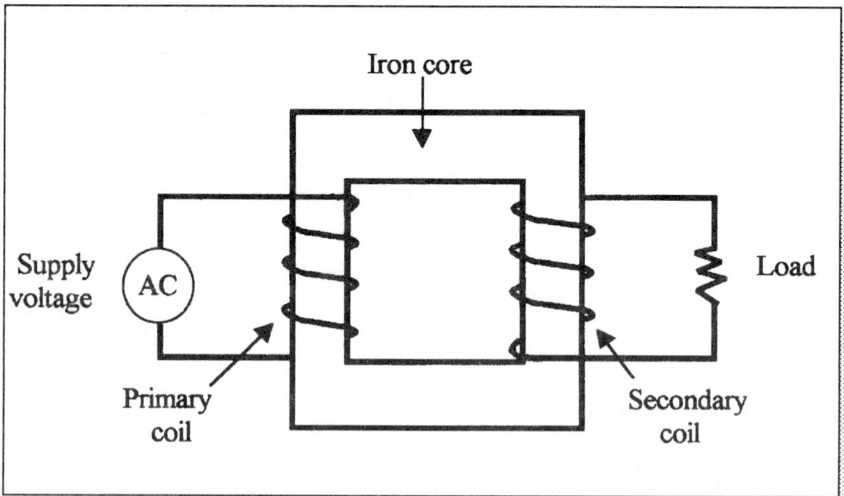

Figure 11.14
Basic transformer.

to another coil, even though there is no electrical connection between them. The number of lines of force available in the primary is determined by the primary voltage and the number of turns on the primary—each turn producing a given number of lines. If there are many turns on the secondary, each line of force will cut many turns of wire and induce a high voltage. If the secondary contains only a few turns, there will be few cuttings and low induced voltage. The secondary voltage, then, depends on the number of secondary turns as compared with the number of primary turns. If the secondary has twice as many turns as the primary, the secondary voltage will be twice as large as the primary voltage. If the secondary has half as many turns as the primary, the secondary voltage will be one-half as large as the primary voltage.

IMPORTANT

Important Point: *The voltage on the coils of a transformer is directly proportional to the number of turns on the coils.*

A voltage ratio of 1:4 (read as "1 to 4") means that for each volt on the primary, there are 4 volts on the secondary. This is called a **step-up** transformer. A step-up transformer receives a low voltage on the primary and delivers a high voltage from the secondary. A voltage ratio of 4:1 (read as "4 to 1") means that for 4 volts on the primary, there is only 1 volt on the secondary. This is called a **step-down** transformer. A step-down transformer receives a high voltage on the primary and delivers a low voltage from the secondary.

11.5 POWER DISTRIBUTION SYSTEM PROTECTION

Interruptions are very rare in a power distribution system that has been properly designed. Still, protective devices are necessary because of the load diversity. Most installations are quite complex. In addition, externally caused variations might overload them or endanger personnel.

Figure 11.15 shows the general relationship between protective devices and different components of a complete system. Each part of the circuit has its own protective device or devices that protect not only the

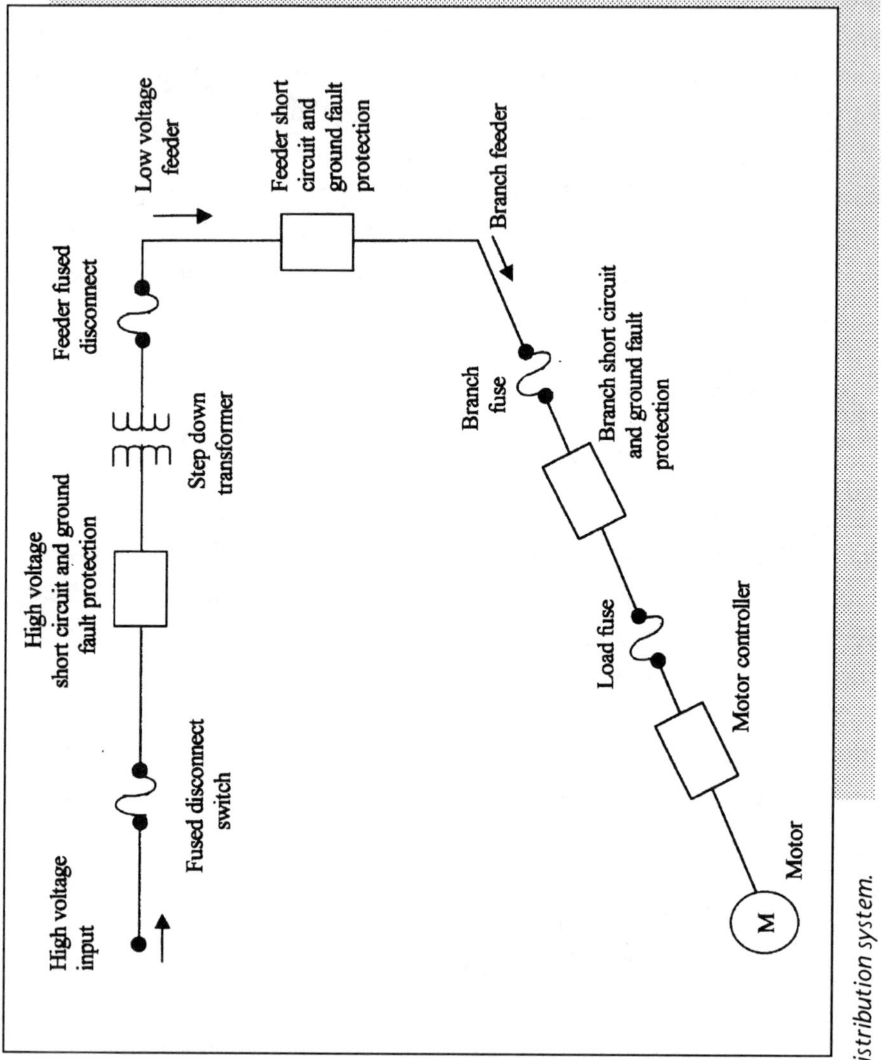

Figure 11.15
Motor power distribution system.

load, but also the wiring and control devices themselves. These disconnect and protective devices are described in the following sections.

11.5.1 FUSES

The passage of an electric current produces heat. The larger the current, the more heat is produced. In order to prevent large currents from accidentally flowing through expensive apparatus and burning it up, a *fuse* is placed directly into the circuit, as in Figure 11.15, so as to form a part of the circuit through which all the current must flow.

IMPORTANT ***Key Point:*** *A fuse is a thin strip of easily melted material. It protects a circuit from large currents by melting quickly, thereby breaking the circuit.*

The fuse will permit currents smaller than the fuse value to flow but will melt and, therefore, break the circuit if a larger, dangerous current ever appears. For instance, a dangerously large current will flow when a "short circuit" occurs. A short circuit is usually caused by an accidental connection between two points in a circuit that offer very little resistance to the flow of electrons. If the resistance is small, there will be nothing to stop the flow of the current, and the current will increase enormously. The resulting heat generated might cause a fire. However, if the circuit is protected by a fuse, the heat caused by the short-circuit current will melt the fuse wire, thus breaking the circuit and reducing the current to zero.

Fuses are rated by the number of amps of current that can flow through them before they melt and break the circuit. Thus, we have 10-, 15-, 20-, 30-A, etc., fuses. We must be careful that any fuse inserted in a circuit be rated low enough to melt or blow before the apparatus is damaged. For example, in a plant building wired to carry a current of 10-A it is best to use a fuse no larger than 10-A so that a current larger than 10-A could never flow.

Some equipment, such as an electric motor, requires more current during starting than for normal running. Thus, fast-time or medium-time fuse rating that will give running protection might blow during the initial period when high starting current is required. **Delayed action** fuses are used to handle these situations.

11.5.2 CIRCUIT BREAKERS

Circuit breakers are protective devices that open automatically at a preset ampere rating to interrupt an overload or short circuit. Unlike fuses, they do not require replacement when they are activated. They are simply reset to restore power after the overload has been cleared.

IMPORTANT ***Key Point:*** *A circuit breaker is designed to break the circuit and stop the current flow when the current exceeds a predetermined value.*

Circuit breakers are made in both plug-in and bolt-on designs. Plug-in breakers are used in load centers. Bolt-ons are used in panelboards and exclusively for high interrupting current applications.

Circuit breakers are rated according to current and voltage, as well as short circuit interrupting current. A single handle opens or closes contacts between two or more conductors. Breakers are single-pole, but can be ganged single-pole units—forming double- or triple-pole devices opened with a single handle.

Several types of circuit breakers are commonly used. They may be thermal, magnetic, or a combination of the two. Thermal breakers are tripped when the temperature rises because of heat created by the over-current condition. Bimetallic strips provide the time delay for overload protection. Magnetic breakers operate on the principle that a sudden current rise creates enough magnetic field to turn an armature, tripping the breaker and opening the circuit. Magnetic breakers provide the instantaneous action needed for short circuit protection. Thermal-magnetic breakers combine features of both types of breakers.

Magnetic breakers are also used in circumstances where ambient temperature might adversely affect the action of a thermal breaker.

An important feature of the circuit breaker is its arc chutes, which enable the breaker to extinguish very hot arcs harmlessly. Some circuit breakers must be reset by hand, while others reset themselves automatically. When the circuit breaker is reset, if the overload condition still exists, the circuit breaker will trip again to prevent damage to the circuit.

11.5.3 CONTROL DEVICES

Control devices are those electrical accessories (switches and relays) that govern the power delivered to any electrical load.

In its simplest form, the control applies voltage to, or removes it from, a single load. In more complex control systems, the initial switch may set into action other control devices (relays) that govern motor speeds, servomechanisms, temperatures, and numerous other equipment. In fact, all electrical systems and equipment are controlled in some manner by one or more controls. A controller is a device or group of devices that serve to govern, in some predetermined manner, the device to which they are connected.

In large electrical systems, it is necessary to have a variety of controls for operation of the equipment. These controls range from simple pushbuttons to heavy-duty contactors that are designed to control the operation of large motors. The pushbutton is manually operated while a contactor is electrically operated.

Self-Test

11.1 Another name for an a-c generator is_______________.

11.2 What is electromagnetic induction?

11.3 An alternator converts _______________ energy into _______________ energy.

11.4 As the armature of an alternator moves through one complete rotation, what must occur for maximum voltage to be generated?

11.5 A transformer transmits energy from one circuit to another by means of _______________ _______________ between two coils.

11.6 In a transformer, the _______________ coil is connected to the source of supply. The _______________ coil is connected to the _______________.

11.7 A step-up transformer _______________ the voltage and _______________ the current. A step-down transformer _______________ the voltage and _______________ the current.

11.8 What is the purpose of a fuse?

11.9 The transfer of electric energy from one coil to another is known as_______________.

11.10 How are circuit breakers rated?

Final Review Examination

The following questions are designed to test your overall understanding of the material presented in this manual.

12.1 Is the charge of an electron positive or negative?

12.2 Which have more free electrons, metal or insulators?

12.3 Which is a semiconductor, copper, silicon, or neon?

12.4 How many electron charges are there in the practical unit of 1 coulomb?

12.5 How much potential difference is there between two identical charges?

12.6 Define "difference in potential."

12.7 The flow of 2 coulombs/second of electron charges is how many amps of current?

12.8 Which has more resistance, carbon or copper?

12.9 The same voltage applied, which resistance will allow more current, 4.9 Ω or 6000 Ω?

12.10 Calculate R for 12 volts and 0.006 A.

12.11 An electron heater takes 15 A from the 120-volt power line. Calculate the power.

12.12 How much is the R of a 100-watt 120-volt light bulb?

12.13 A series circuit has IR drops of 10 volts, 20 volts, and 30 volts. How much is the applied voltage E_T of the source?

12.14 Each of three equal resistances dissipates 2 watts. How much P_T is supplied by the source?

12.15 Parallel branch currents are 1 A for I_1, 2 A for I_2, and 3 A for I_3. Calculate I_T.

12.16 Find R_T for three 4.7-MΩ resistances in parallel.

12.17 How much is the *IR* voltage drop across a closed switch?

12.18 How much is the resistance of a good fuse?

12.19 Which type can be recharged, a primary or a secondary cell?

12.20 An a-c voltage varies in ________________ and reverses in ________________.

12.21 A sine-wave voltage has a peak value of 170 volts. What is its value at 45°?

12.22 State the law of attraction and repulsion of charged bodies.

12.23 What is true power?

12.24 In a series circuit whose source voltage is 30 volts, there are three resistors. The voltages across two of the resistors are 6 volts and 10 volts. What is the voltage across the third resistor?

12.25 Explain the difference between d-c and a-c current?

12.26 What is the *L/R* time constant of a circuit whose total inductance is 100 mh and whose total resistance is 50 Ω ?

12.27 How much Z_T for a 20-Ω *R* in series with a 20-Ω X_C?

12.28 How much is the *RC* time constant for 470 pF in series with 2 MΩ on charge?

12.29 What is the unit for apparent power?

12.30 In a capacitor, is the electric charge stored in the dielectric or in the metal plates?

Appendix A

Answers to Chapter Self-Tests

CHAPTER 1

1.1 Resistance = 1/5 mho = 0.2 ohms

1.2 Orbits or shells

1.3 Protons, neutrons

1.4 Atomic number

1.5 Attract, repel

1.6 The force that causes free electrons to move in a conductor as electric current

1.7 Friction, pressure, heat, light, chemistry, magnetism

1.8 (a) Volt, *E*; (b) ampere, *I*; (c) coulomb, *Q*; (d) ohm, *R*

1.9 The value of the resistors, the length of the conductors, and the diameter of the conductors

1.10 Direct current flow does not change direction, while alternating current periodically changes direction.

1.11 The (magnetic) poles

1.12 The flux lines, or magnetic flux, along which a magnetic force acts

1.13 Natural magnets, permanent magnets, and electromagnets

1.14 Chemistry; magnetism

1.15 Voltage and resistance

CHAPTER 2

2.1 Chemistry

2.2 Battery, two

2.3 Negative

2.4 Opposite

2.5 The electrodes and the electrolyte

2.6 Secondary

2.7 Parallel

2.8 The form of the electrolyte

2.9 Series-parallel

2.10 A moist paste

CHAPTER 3

3.1 A series circuit has only one path for current flow, while a parallel circuit has more than one path.

3.2 Source voltage

3.3 Voltage drop

3.4 Counterclockwise

3.5 $I_T = E/R$
$= 40/20$
$= 2$ amp

3.6 $E = IR_1$
$= 2 \times 6$
$= 12$ V

3.7 $P = I^2 R_2$
$= 2^2(4)$
$= 4 \times 4$
$= 16$ V

3.8 $P = EI$
$= 40 \times 2$
$= 80$ watts

3.9 Less, more

3.10 Zero, excessive

3.11 120 ohms

3.12 One

3.13 Individual, total

3.14
$$\frac{1}{R_T} = \frac{1}{16} + \frac{1}{32} + \frac{1}{160}$$

$$\frac{1}{R_T} = \frac{10}{160} + \frac{5}{160} + \frac{1}{160} = \frac{16}{160}$$

$$R_T = \frac{160}{16} = 10 \text{ ohms}$$

3.15
$$I_T = \frac{V}{R_1} + \frac{V}{R_2} + \frac{V}{R_3}$$

$$= \frac{20}{16} + \frac{20}{32} + \frac{20}{160}$$

$$= 1.25 + 0.625 + 0.125$$

$$= 2.000 \text{ amps}$$

3.16 $P_T = 20\text{V} \times 2$ amps
$P_T = 40$ watts

3.17 12.24 ohms

3.18 1 amp (rounded)

3.19 1.5 V

3.20 12 watts (rounded)

CHAPTER 4

4.1 To convert a source of applied voltage to a load resistance with a minimum *IR* voltage drop in the conductor so that most of the applied voltage can produce current in the load resistance.

4.2 Resistivity

4.3 Circular mil

4.4 Circular mil

4.5 Conductivity

4.6 Smaller

4.7 Doubles

4.8 Greater, any

4.9 Allowable power loss in the line; permissible voltage drop in the line; the current-carrying capacity of the line; conductor installed where ambient temperature is relatively high

4.10 Negative

4.11 The ability of the insulator to withstand potential difference, it is usually expressed in terms of the voltage at which the insulation fails because of the electrostatic stress.

4.12 High

4.13 Original

4.14 It will withstand high voltages.

4.15 From Table 4.2, the cross-sectional area for No. 20 wire is 1020 CM. From Table 4.1, ρ for copper is 10.37 CM (Ω/ft). Use Equation (4.3) to find the resistance of 500 ft of wire.

$$R = \rho \frac{L}{A} = (10.37) \times 500/1020 = 5.08\ \Omega$$

CHAPTER 5

5.1 The two are directly proportional. As flux density increases, field strength also increases.

5.2 Grasp the conductor in the left hand so that the thumb points in the direction of current flow. The fingers will curve around the conductor in the direction of the magnetic field.

5.3 The type of material and the flux density

5.4 The lag between magnetization and the forces that produce it

5.5 a. Increase the number of turns of wire in the coil
b. Increase the current flow through the coil
c. Select a core material with higher permeability.

5.6 When current flows in the coil

5.7 Permeability

5.8 Field intensity

5.9 At = mmf

5.10 North pole

CHAPTER 6

6.1 The conductor(s) must cut a maximum number of lines of force.

6.2 Will increase

6.3 141.4 volts (200 X 0.707)

6.4 A voltage is induced in the conductor.

6.5 It periodically changes direction.

6.6 A waveform that represents all the instantaneous values of current or voltage as the armature of an alternator moves through one complete rotation.

6.7 Magnitude and direction

6.8 846 volts (600 volts X 1.41)

6.9 The number of sine waves generated each second

6.10 Speed of rotation of the armature; the number of poles producing the magnetic field

CHAPTER 7

7.1 A-c, cut, counter

7.2 Counter

7.3 Current has an associated magnetic field

7.4 When current flows in the conductor, the associated magnetic field moves across (cuts) the conductor, inducing an emf.

7.5 Number of turns in the coil; length of the coil; cross-sectional area of the coil; permeability of the core

7.6 Increase

7.7 Decrease

7.8 4 milliseconds

7.9 Increase

7.10 Increase

CHAPTER 8

8.1 Has two metal plates separated by an insulating material (dielectric)

8.2 One millionth

8.3 Area of plates, distance between plates, dielectric constant

8.4 The time required to charge a capacitor to 63% of its maximum value or to discharge it to 37% of its maximum value.

8.5 120 μf

8.6 Dielectric

8.7 They are polarity sensitive.

8.8 Charge

8.9 Electrons

8.10 Increases, decreases

CHAPTER 9

9.1 A-c

9.2 Increases

9.3 Inductance; frequency; 2π

9.4 Capacitance; frequency; 2π

9.5 $X_L = 6.28 \times 500 \text{ Hz} \times .03 = 94.2\,\Omega$

9.6

$$X_C = \frac{1}{6.28 \times 200 \text{ Hz} \times 0.00012 \text{ f}} = 6.63\,\Omega$$

9.7 Voltage leads current by 90°

9.8 Voltage lags current by 90°

9.9 Impedance

9.10 Average power actually consumed by an a-c circuit

CHAPTER 10

10.1 Average power

10.2 Apparent power

10.3 90

10.4 0

10.5 0

10.6 1

10.7 Effective values and phase angles

10.8 When $X_L = X_C$

10.9 Ratio of average power to apparent power

10.10 *R* (resistance)

CHAPTER 11

11.1 Alternator

11.2 The effect that causes current flow in a conductor moving across magnetic lines of force

11.3 Mechanical; electrical

11.4 The conductor(s) must cut a maximum number of lines of force

11.5 Electromagnetic induction

11.6 Primary; secondary; load

11.7 Increases; decreases; decreases; increases

11.8 To protect an electrical circuit and load.

11.9 Magnetic coupling

11.10 According to current and voltage

Appendix B

Answers to Chapter 12—Final Review Examination

12.1 Negative

12.2 Metals

12.3 Silicon

12.4 6.25×10^{18}

12.5 Zero

12.6 The force that causes free electrons to move in a conductor as electric current

12.7 2-A

12.8 Carbon

12.9 4.9 Ω

12.10 2000 Ω

12.11 $P = EI = 120\text{ V} \times 15\text{ A} = 1800\text{ w} = 1.8\text{ kw}$

12.12 $$R = \frac{E_2}{P} = \frac{(120\text{ V})^2}{100\text{ w}} = 144\ \Omega$$

12.13 60 volts

12.14 6 watts

12.15 $I_T = 6$ A

12.16 $R_T = \dfrac{R}{N} = \dfrac{4.7}{3} = 1.57\ \text{M}\Omega$

12.17 Zero

12.18 Zero

12.19 Secondary

12.20 Magnitude; polarity

12.21 170 V x cos 45° = 120.2 V

12.22 Like charges repel each other and unlike charges attract each other.

12.23 Power that is dissipated as heat

12.24 $V_T = V_1 + V_2 + V_3$

$30\ \text{V} = 6\ \text{V} + 10\ \text{V} + \text{V}_3$

$V_3 = 14$ V

12.25 D-c flows in only one direction, while a-c periodically reverses the direction of flow.

12.26 2 milliseconds (divide 100 mh by 50 Ω)

12.27 $Z_T = \sqrt{R^2 + X_C^2} = \sqrt{20^2 = 20^2} = 28.28$

12.28 940 μs

12.29 Voltamp

12.30 Dielectric

Index